MIC = Minimum Inhibitory
Concentration of drug
needed to inhibit bacterial
growth. — The higher the MIC,
the greater the resistance.
A bacterium is defined as
susceptible if the MIC
$\leq$ 4 mg/L   (pp 72-73)

THE
KILLERS WITHIN

THE
KILLERS WITHIN

THE DEADLY RISE OF
DRUG-RESISTANT BACTERIA

MICHAEL SHNAYERSON
MARK J. PLOTKIN

LITTLE, BROWN AND COMPANY
BOSTON NEW YORK LONDON

FIRST EDITION

Library of Congress Cataloging-in-Publication Data
Shnayerson, Michael.
    The killers within : the deadly rise of drug-resistant bacteria /
Michael Shnayerson, Mark J. Plotkin.
        p.   cm.
    Includes index.
    ISBN 0-316-71331-7 (hc)
    1. Drug resistance in microorganisms—Popular works.   I. Plotkin, Mark J.
II. Title.
QR177 .S43   2002
616'.01—dc21   2002024177

10 9 8 7 6 5 4 3 2 1

Q-MART

Text design by Stanley S. Drate/Folio Graphics Co. Inc.

Printed in the United States of America

This book is for our daughters—
Ann Lauren, Gabrielle, and Jenna—
with the hope that they don't grow up to live
in a post-antibiotic world.

CONTENTS

THE
KILLERS WITHIN

PROLOGUE

D r. Glenn Morris was growing very worried. His patient was not supposed to die.

The son of Southern Baptist missionaries, Morris had grown up in Bangkok, Thailand. There he had witnessed firsthand the magnitude of devastation that could be wrought by a bacterial plague. During the dreaded dry season after the monsoons had passed, waves of cholera would sweep through the canal-laced city, killing hundreds at a time. Morris would never forget the screams of ambulances racing through the streets. Cholera seemed to strike without warning: a man who'd sampled the food from a street vendor would be hideously sick just hours later, lying limp and helpless as his vital nutrients flowed out of him amid ceaseless diarrhea. Death would often follow. Was it the food? The water? Who knew? Morris had heard about the horrors of hell in church on Sunday. He didn't think it could be much worse than the dread permeating a city in the grip of a cholera epidemic.

Morris's parents had assured the shaken child that by staying away from street food, drinking boiled water, and, most important, taking fluids and antibiotics at the slightest cholera-like symptom, he would be safe. But as Morris now looked down at his suffering patient, he knew that such soothing assurances had no relevance here.

A year before, Ed Burke[1] had taken his good health for granted. Forty years old, lean and fit, he was a recently divorced accountant living with his mother while he tried to put his life together again. But Burke had been feeling weak and tired when he went to the University of Maryland's Medical Center for a checkup. He told doctors he'd been having stomach pains and chronic colds. A routine blood test revealed that he was suffering from leukemia. Though Burke was shocked and frightened by the diagnosis of cancer, his physician explained that most forms of leukemia responded well to chemotherapy. In all likelihood, he'd be able to undergo the regimen and soon resume a normal life.

That was the beginning of the end.

Burke's physician initiated chemotherapy almost immediately. Though often effective against leukemia and other cancers, the drastic treatment—with its searing chemicals that course through the body like Drano—can have the undesired effect of suppressing the immune system as well, sometimes leading to bacterial infections that the weakened immune system cannot contain. Physicians use antibiotics to help eradicate these potentially life-threatening infections. Sometimes these bugs prove resistant to the initial antibiotic, in which case the physician simply switches to another one. For decades, there had always been plenty in reserve. But for this particular case, the reserve had been exhausted.

Burke's infection was caused by the bacterium known as *Enterococcus faecium*. One expert calls *E. faecium* the cockroach of microbial pathogens: proliferating freely in the gastrointestinal tract, it usually causes no more trouble than roaches colonizing a dark cupboard. But when breakdowns in the immune system allow the bugs to escape, they begin to cause serious infections, anywhere from the heart down to the urinary tract. After proving resistant to the initial

---

[1]Some names, including this one, have been changed to protect the privacy of patients and their families.

antibiotics used, Burke's *E. faecium* also showed resistance to vancomycin, an older but still powerful antibiotic that represented the last-chance treatment for resistant enterococci when all else failed. This time vancomycin failed as well: vancomycin-resistant *E. faecium,* better known as VRE, had appeared in Burke's bloodstream, a dangerous escalation. That was when Morris had been called in.

With his stocky build and powerful arms and shoulders, the ruddy-faced Morris looked more like an ex-linebacker than a man of medicine, though his soft-spoken manner made him, at first glance, seem shy and retiring. But when he was stationed behind a microphone at a scientific conference and given a rapt audience for his impassioned calls to action about disturbing bacterial advances, the Southern preacher inside him very clearly emerged.

Awed by the bacterial devastation he had witnessed in Thailand, Morris had grown up determined to do what he could to prevent such suffering in the future. Through willpower, hard work, and a keen intellect, he had turned himself into a formidable microbe hunter: schooled in tropical medicine, public health, food safety, and genetics, he had a breadth of training possessed by few in his field. Now Head of Infectious Diseases at the University of Maryland's Baltimore Veterans Affairs Medical Center, Morris was one of the country's best-known experts in his field. And Maryland needed an expert. In the last few years, Morris had seen an explosive growth of VRE right in Baltimore.

The full import of this trend was difficult for patients to absorb. Certain strains of *E. faecium* were resistant to nearly all of the more than one hundred antibiotics that modern science had produced. They were, quite simply, unstoppable.

Even harder to explain to patients was that *E. faecium* was a hospital bug. Almost certainly, it had infected Burke *after* he was in the care of his oncologists. *It had infected him right there in his hospital bed.* And how had it gotten there? Probably by alighting from the unwashed hands of a busy doctor, nurse, or other healthcare worker who had just had contact with another patient carrying the bug. Oncology wards and intensive care units of nearly all hospitals were notoriously rife with resistant bugs, though few institutions would admit as much. All too often, these drug-resistant bugs took weeks to develop into infections, so that the doctors and nurses who had inadvertently passed them to a patient might not ever learn what they had done.

For elderly patients with chronic illnesses and ravaged immune systems, VRE was proving lethal, the extra infection that nudged an already sick person over the edge. Younger patients were usually impervious to it—unless, that is, their own immune systems were compromised by chemotherapy, as Burke's was, or by drugs given to prevent rejection of a transplanted kidney, or by some life-threatening, out-of-the-blue calamity: a car accident, perhaps, or a third-degree burn. Then they were as susceptible as patients twice their age. Given that well over one million Americans were diagnosed each year with some form of invasive cancer—15 million since 1990—the number of potential victims for VRE was surprisingly large.

When Morris entered Burke's hospital room the first time, the accountant had looked up at him with desperate hope. Gently, Morris had had to explain that he had no magic bullet for Burke's VRE—no cure at all. He could only hope that with the end of chemotherapy, Burke's white blood cell count would bounce back up quickly enough for his immune system to handle the infection itself.

As if in answer to the Burke family's prayers, that was what happened—at first. The leukemia disappeared—whether it was in remission or gone for good, the doctors could not yet tell—and chemotherapy was halted. Burke's immune system began to recover and started producing the white blood cells responsible for killing bacteria that invade the bloodstream. As the white blood cells attacked the infection, the patient's fever broke, and he felt stronger every day. Within a week, he was sent home.

Just ten months later, Burke was back in the hospital with a relapse of leukemia. Reluctantly, his doctors gave him more chemotherapy. When they did, VRE reappeared in his bloodstream. It had been lurking in his intestinal tract, a killer within. Twice more the man recovered enough for his immune system to fight the VRE to a standoff. But the superbug was not yet beaten.

A year later, another relapse of leukemia, and more chemotherapy, pushed Burke's white blood cell count down too far. The bug that had been his constant companion, as Morris grimly put it, once again infected his bloodstream. Back came the high fever, the chills, the irregular heartbeat, the shortage of breath. But this time there was no rebound, even when chemotherapy was halted. Burke began vomiting, and his blood pressure plunged. As the flow of blood to his brain slowed to a trickle, his vision dimmed and he became disoriented. At the same time, his ever weakening heartbeat pumped less and less blood to his other vital organs. One by one they began shutting down, like lights winking out during a power blackout. As the kidneys and liver ceased to operate and cleanse his body of waste materials, Burke essentially poisoned himself. Finally came full-blown septic shock. Burke went pale and delirious, cold and clammy to the touch. He suffered a series of small heart attacks, and began to suffocate as his lungs filled with fluid. Eight days after the VRE infected his bloodstream a final time, Burke succumbed.

That was in September 1995.

It doesn't get much worse than this, Morris thought at the time: a forty-year-old man with an infection no antibiotic can stop. But he would be wrong. Over the next six years, a grim new era of multi-resistant bacteria would unfold in—and out of—hospitals around the globe, making Burke's case seem all too typical. Relentlessly, these newly hardy, invisible bugs would proliferate all around us, some festering on bedrails and seat cushions, telephones and thermometers, others passing through the air from one human host to the next. Silently, they would colonize even the healthiest of us, coating our skin, nestling in our noses, spreading in our throats, swimming through our stomachs and gastrointestinal tracts—until it could not be said that any of us was ever without at least a smattering of highly drug-resistant bugs, waiting for the chance to infect those among us who grew suddenly weak and sick.

The bugs were everywhere, exponentially multiplying. And each year now, fewer drugs seemed able to stop them.

1

THE SILENT WAR

Most mornings for Glenn Morris started with his daughters. Only after he loaded the three of them—aged fifteen, twelve, and nine—into his old Infiniti G20 and dropped them off at the carpool did he head in to the hospital. But on the mornings of July 2001, while the girls were on summer vacation, Morris bid his wife goodbye and drove off alone to the front lines of a war none of his neighbors could see or hear.

As a doctor in his late forties who was both head of epidemiology at his hospital and chairman of the associated university department, Morris could have graduated from going on clinical rounds. Still, he made a point of doing it two months a year. You couldn't just teach and do research, he believed—you had to see what new infections patients were incurring. Also, going on rounds made him feel the same stomach-tightening anticipation of the unknown that he'd experienced as a medical resident more than two decades before. And so he headed in from his Tudor-style house in Roland Park—a leafy neighborhood of large, comfortable homes built a century ago as one of Baltimore's first suburbs—to spend his days treating half a hundred very sick patients, many of them indigent, in the general ward of Baltimore's Veterans Affairs Medical Center.

On the fifteen-minute drive into the city, Morris liked to listen to country-and-western music, its trucks and trains and broken hearts weaving through his thoughts of doxycycline or ciprofloxacin for one patient, vancomycin or Synercid for another. After parking in the hospital's underground garage and ascending, white-jacketed, to the general ward on the third floor, he started by checking his charts. Six new patients, he saw, had been admitted to the ward by way of the emergency room. One's condition looked especially bad.

Morris went from bed to bed, trailed by a note-taking team of medical students, interns, and residents. Because this was a VA hospital, most of the patients in the general ward were male, elderly, and afflicted with chronic conditions. Many also had symptoms that indicated bacterial infection. A decade ago, antibiotics[1] would have knocked out all of these infections almost immediately. Now on average, about 20 percent of patients on Morris's clinical rounds had infections resistant to one, two, three, or more drugs. When he wrote for medical journals, Morris described this multidrug resistance in dry, clinical terms that expressed none of the emotions he felt when he witnessed the ravages of an almost unstoppable infection. What he felt was dismay, and alarm, and a little twitching of fear.

When Morris pointed out antibiotic-resistant infections to his interns and residents, he didn't need to emphasize that these were *bacterial* infections. They'd had it drilled into them in medical school that most infections are either bacterial or viral, and that bacterial infections are the ones that respond to antibiotics.[2] Viruses, they knew, were a whole other matter. A virus is a tiny squiggle of protein-covered

[1]Strictly speaking, the term *antibiotics* refers only to substances found in nature. With the advent of synthetic drugs, the term *antimicrobials* has evolved to encompass both natural and synthetic drugs. So rooted is the term *antibiotic* in both medical and lay literature, however, that even doctors commonly use it to refer to either natural or human-engineered drugs. For simplicity's sake, the term as used in this book will be given this broader context.

[2]Some infections are also caused by parasites (malaria and tapeworm, for example), which are uncommon in the United States and not associated with hospital infections; others are caused by fungi, some of which do plague U.S. hospital patients (among them Candida and Aspergillus). But bacteria cause far more mortalities in U.S. hospitals than parasites or fungi.

DNA or RNA, so small it isn't even a living, cellular organism: its only function is to bore into the cells of other organisms and force those cells to produce more viruses. (AIDS is caused by a virus; so is the common cold.) Antibiotics are useless against viruses. Bacteria, on the other hand, are one-celled organisms: the smallest creatures on the planet. The cell has various parts that enable the bacterium to live and replicate. Those parts can be targets for antibiotics. Unless, that is, the bacteria figure out how to change or deflect the drugs and make themselves resistant.

A decade ago, Morris liked to remind his entourage, doctors had only to reach for penicillin, or one of the third-generation cephalosporins, or the then new, brilliantly effective fluoroquinolones. Now for empiric therapy—immediate treatment of new patients, before a lab could determine exactly what bug they had—doctors often found themselves in the dark, guessing which antibiotic would work. Often there was time to correct the therapy once cultures provided a profile of which drugs still worked against a bug. Sometimes there wasn't. Whenever a newspaper obituary listed cause of death as "complications" following surgery, chances were that a doctor had guessed wrong in terms of antibiotics—or that a bug had proved resistant to all of them. This was code that all healthcare workers, hospital staff, and HMO providers understood but few outside the medical world knew.

Most at risk were the old and the infirm, their immune systems deteriorated, especially in hospitals: at the dawn of the twenty-first century, roughly a third of all people older than sixty-five were dying from infections. Nearly as vulnerable, however, were the very young. Their immune systems were immature, not ravaged, but the result was the same. Tough, sometimes unstoppable strains of the usual suspects—especially *Streptococcus pneumoniae*—caused terrible, recurrent ear infections, or meningitis, or systemic bloodstream infections that shut down a child's vital organs. Every year, 1.2 million children around the world were estimated to die of *S. pneumo,* the leading bacterial cause of pneumonia. In the United States alone, *S. pneumo* was said to cause 500,000 cases of pneumonia, many of them pediatric, as well as 7 million ear infections, most of them pediatric, too. Only a decade ago, nearly all strains of *S. pneumo* had been susceptible to penicillin, the drug of choice for these infections. Now 45 percent of all *S. pneumo* strains were penicillin resistant. Some skeptics observed that with *S. pneumo,* a doctor could increase the dose of antibiotics and still hope to prevail in many cases. But that was cold comfort to parents who saw their children's lives imperiled. Gary Doern, Director of Clinical Microbiology at the University of Iowa Hospital in Iowa City, tracked *S. pneumo* on a national, ongoing basis and was staggered by its fast-rising rates of resistance. "Do the math," he said grimly. "Where will it be fifteen years from now?"

S. pneumo claimed as many victims outside the hospital as it did because, unlike many bacterial pathogens, it was spread by droplets: coughing passed it from host to host. *Enterococcus faecium* and *Staphylococcus aureus* infected hospital patients for the most part. But with *S. aureus,* the most virulent of the three, there were signs that that was changing.

In January 2001, Bryan Alexander, eighteen, was found guilty of assault and drunken driving and sentenced to a 180-day term at a correctional boot camp in Mansfield, Texas. On January 4, he filed a written request for medical attention. According to his father, he filed two more requests; all three requested treatment at the local hospital. The camp nurse chose to refuse them. On January 9, Alexander died of pneumonia caused by a *S. aureus* infection: an otherwise healthy eighteen-year-old killed by microscopic organisms in just days. A few months later, talk show host Rosie O'Donnell very nearly died after cutting her finger with a fishing knife and incurring a multidrug-resistant *S. aureus* infection. "On Tuesday night, April 3 [2001], my hand started to hurt. A lot. It was an itchy-hot-burning-searing-what-the-hell-is-happening pain," she recalled. The pain became unbearable; by the next day, O'Donnell was in the hospital, her hand so swollen it looked "like a kid's bright-red baseball mitt." Multiple surgeries were needed to debride her finger—to cut away the dead and infected tissue—and decontaminate the site. Neither good health nor celebrity had protected these victims.

Strains of all three of these common bacterial infections—*E. faecalis,* *S. aureus,* and *S. pneumo*—were now multidrug-resistant and spreading into the community. Strains of other bacteria—*Acinetobacter baumannii, Pseudomonas aeruginosa,* and *E. faecium*—remained hospital-bound but had become resistant to *all* antibiotics. So widely and quickly were bacteria of different species trading their resistance genes that the vast, invisible world of bacteria could be thought of as a single, miasmic, multicelled organism, its trillions of parts all working together for the common goal of survival against antibiotics. What this boded for humans, the bugs' primary source of food, was in no way good.

At the bedside of the patient whose case history worried him the most, Morris offered greetings with a cheer he didn't feel. The patient, a man in his seventies, had come to the hospital some time ago for a routine knee replacement. Apparently, while his knee was cut open in surgery, he'd incurred a methicillin-resistant *S. aureus* infection, or MRSA. Nearly all strains of *S. aureus* were now resistant to penicillin; almost half the hospital strains were also resistant to methicillin, the drug once thought to be a permanent replacement for penicillin. The infection had manifested itself a month after the man was back home. In he came again to the hospital for surgery to decontaminate the joint, followed by a six-week course, also at the hospital, of vancomycin.

Vancomycin was a last resort, but that didn't make it a great drug. It often failed to penetrate deep bone infections, and it had to be administered intravenously, which meant using catheters, which became conduits for other disease-causing, or pathogenic, bugs. In this case, when vancomycin failed to stem the infection, the man's doctors removed the artificial joint altogether and fused the joint that remained. Then they hit him with another six-week course of vancomycin. Now he was back again, this time with a fever that almost certainly signaled the return, yet again, of his resistant infection. He had bedsores, a urinary catheter, a fused knee that was essentially worthless, and deep infections that just wouldn't quit. He was almost pathologically depressed, as well. His wife had remained a constant presence at his hospital bedside, but she was on the verge of a breakdown herself, unsure whether the downward spiral of complication after complication could ever be reversed.

Morris knew he had to prescribe vancomycin. He had no choice. But where to put the IV? The man had endured so many intravenous lines he was running out of veins. Reluctantly, Morris put him on vancomycin via a central line—a catheter introduced into one of his large veins—and wished him luck. Privately, Morris thought the man would be lucky to live out the year.

This was a case, Morris thought, that should never have happened: a man who'd come into the hospital in basically good health and emerged with a dire strain of MRSA. Doctors had a phrase they used among themselves to refer to such patients—the ones with infections resistant to one or more drugs and who seemed too sick to respond to any antibiotics.

Train wrecks, they called them.

Not every doctor and microbiologist at the dawn of the twenty-first century felt, as Morris did, that the golden era of antibiotics might be coming to an end. Not all felt that bacterial resistance had become, in the words of one physician, one of the greatest threats to the survival of the human species. But many did. And all agreed that resistance had become an urgent global issue. Stuart Levy, M.D., a Tufts University professor whose Cassandra-like warnings on the subject two decades before had all come to pass, saw only worse things to come. "We are clearly in a public health crisis," he said to anyone who would listen. "In fact, we're on the road to an impending public health disaster." Joshua Lederberg, Ph.D., Nobel laureate and longtime leading expert in antibiotic resistance at New York's Rockefeller University, felt that by comparison, the Ebola virus was small potatoes. "The odds of Ebola breaking out are quite low, but the stakes are very high. With antibiotic resistance, the odds are certain and the stakes are just as high. It is happening right under our noses."

The principal cause was overuse—and misuse—of antibiotics. In 1954, 2 million pounds of antibiotics had been produced in the United

States. By the end of the century, the annual figure had risen, by some estimates, to more than 50 million pounds. Yet researchers at the federal Centers for Disease Control and Prevention (CDC) in Atlanta, Georgia, judged that a full third of the 150 million outpatient prescriptions for antibiotics written each year in the United States were unnecessary: either the infection turned out to be viral or the wrong drug was prescribed. Doctors prescribed the drugs partly to placate demanding patients and partly to protect themselves legally if they failed to prescribe an antibiotic for an infection that turned out to be direly bacterial. The proliferation of antibiotics killed many bacteria but gave the hardiest few some more chances to learn how the drugs worked—and how to resist them.

It was a phenomenon that biologists called selective pressure. Among the billions of bacteria in a drop of human blood, or on a pinpoint of skin, or in a minute isolate of phlegm in the throat or stomach acid, might be a few—just a few—with a chance mutation that enabled them to resist the antibiotic used against them. If the antibiotic was then removed because the patient felt better and stopped using it—or sometimes even if it wasn't—those few resistant bugs would have an ecological niche, or clear field, in which to run wild. Because bacteria replicated so quickly—some bugs created a whole new generation every twenty minutes—the mutants could soon fill the niche. The pressure of the antibiotic, rather than obliterating them, had selected them to survive.

Only a small portion of blame could be pinned on doctors in the community. Lethally resistant bacteria now resided in every hospital and nursing home in the world. Every year in U.S. medical institutions, 2 million patients contracted infections—bacterial, viral, and otherwise—and 90,000 died. Of those 90,000, many had drug-resistant bacterial infections, mostly *S. aureus*. The CDC estimated that 40,000 Americans died each year of those infections.[3] That was

---

[3] The real number was likely higher, because patients with other chronic illnesses, like AIDS or various cancers, might be nudged over the edge by a concurrent drug-resistant infection that wasn't credited as the cause of death.

more than half the number of servicemen who had died during the entire Vietnam War. These deaths occurred in ones and twos, in hospital beds spread across the country, not by the scores on a single battlefield, so they tended to be noticed only by the patients' family and friends; by the hospitals, which certainly did nothing to publicize deaths caused by organisms within their institutions; and by HMOs, which quietly raised their premiums to help cover the estimated $5 billion cost of treating drug-resistant infections each year in the United States. Doctors and researchers published academic papers on drug-resistant bacteria, and, every year, their concerns grew more urgent, their prognoses more bleak. But the public remained largely oblivious to the problem, and as it did, incomprehensibly large populations of bacteria grew more and more resistant to more and more drugs.

Often these resistant bacteria, once established by selective pressure, were passed by contact, on the hands of doctors or nurses, from patient to patient. Many found easy access to their victims' bloodstream through surgical incisions or wounds or by lingering on catheters and prostheses. One study had found a high incidence of pathogenic bacteria on computer keyboards and faucet handles in intensive care units, or ICUs. Another had found the bugs in the cushions and fabric of chairs in hospital common rooms, and in the acoustical tiles of hospital ceilings—lingering there, sometimes, for years. A third had found them on rectal thermometers, a fourth on stethoscopes.

These various reservoirs dramatized the other dimension of the problem. If misuse of antibiotics created drug-resistant bacteria in the first place, poor infection control in hospitals allowed the bugs to spread. Every time a doctor or nurse failed to wash his or her hands before entering a patient's room, millions of invisible pathogens potentially came along for the ride. Yet how feasible was it for emergency department doctors to wash their hands before and after treating each next desperate patient, at a rate of five or six patients an hour? Or for doctors seeing up to two dozen patients on clinical rounds to do the same? In fact, one recent study conducted at Duke

University had determined that only 17 percent of doctors treating patients in an intensive care unit washed their hands thoroughly and consistently. But to the bugs, every patient in an ICU presented another irresistible meal, and, with lax infection control, the bugs got fed.

The most prevalent pathogens were bacteria that people carried with them as part of their natural "flora." In their stomach and intestinal tract milled billions of enterococci. In their throat resided billions more streptococci. In their nose, and on their skin, lived the most worrisome of the big three: staphylococci. Some of these bugs were essential to digestion; others promoted health by staking turf that might otherwise be colonized by more virulent bugs. But given access to a weakened host—often through a cut in the skin—certain strains of these three species could be very bad bugs indeed. Enterococci caused skin and bloodstream infections; under the right circumstances they infected heart valves, too. Streptococci caused all manner of infections, from sore throats and earaches to pneumonia to the horrific necrotizing fasciitis, better known as flesh-eating bacteria. *S. aureus,* the most virulent of the staphylococci, was also alarmingly widespread: between 20–40 percent of people, both healthy and sick, carried *S. aureus,* usually in their nose or on their skin. Once it managed to enter the bloodstream of an immunocompromised person, *S. aureus* caused surgical infections, pneumonia, heart and brain infections, and systemic bloodstream infections that shut down vital organs one by one with an inexorable end result.

In the last decade, the bugs had acquired intricate mechanisms of resistance more quickly, as if the bacterial world was mirroring humanity's own ever quickening pace of development. Some succeeded in making their cell walls impermeable to antibiotics. Others created tiny pumps that actually vomited them out of the cell. Many antibiotics targeted one enzyme or another of the cell wall itself, attaching to it just as the bacterium was making more cell wall enzymes in order to replicate; yet many bugs had figured out how to change or replace those enzymes so that the drug failed to attach. Still other bugs' enzymes attacked the drug itself, slicing its chemical rings. The broad-spectrum antibiotics that most doctors reached for first were the ones likeliest to provoke these mechanisms. They killed a wide range of bugs, as the term implied, but used frequently they also gave that wide range of bugs more chances to develop successful mutations or import resistance genes from other bacteria. As microbiologist Barry Kreiswirth of New York City's Public Health Research Institute put it, "The bugs are getting stronger — and they're getting stronger *faster.*"

Stuart Levy, a puckish fellow given to bow ties and elegant suits, often observed in his lectures, and in his classic book *The Antibiotic Paradox,* that the answer had caused the problem. Or rather, the answer to one problem had led to the next problem. Antibiotics had changed the world, eradicating the horror of pervasive infections that killed young and old alike. They had transformed surgery from a butchery in which most patients died of infections into a modern medical science. Yet the development of novel invasive therapies like organ transplants, prosthetic implants, dialysis machines for kidney failure, and chemotherapy for cancer had resulted in more and more immunosuppressed patients, which in turn provided additional fodder for the microbes. And the better that modern medicine enabled patients to overcome once-lethal conditions like faulty hearts or cancer, the longer it enabled them to live, the more likely they were to decline gently into the clutches of invisible microbial pathogens. "We can close the books on infectious diseases," U.S. Surgeon General William Stewart had declared in 1969, suggesting, in a breathtaking show of hubris, that humans had beaten the bugs once and for all. But the bacteria were fighting back — and gaining on us.

In the early 1990s, only doctors and nurses in hospitals had worried about drug-resistant bacteria. Now, like so many microscopic prisoners, the bacteria were breaking out. They caught their rides on the skin or in the intestinal tract of recovering patients in home care. They clung to aging patients shuttled back and forth between hospitals and long-term care facilities or nursing homes, especially in crowded cities like New York, which had become the epicenter in the United States of drug-resistant bacteria. They migrated to other places where people crowded together: prisons, military barracks, college dormitories, and, most frightening of all, daycare centers. Among the most likely carriers—or vectors, as the literature had it— were the doctors and nurses themselves. Once out, the bugs passed their resistance genes on to other bacteria, and resistance spread exponentially. What rule, after all, had ever restricted resistant bugs to hospitals? No rule *they* knew of.

Resistance flowed from hospitals, it radiated out from antibiotic misuse by doctors in outpatient settings, and it welled up, too, from a third, ubiquitous source in the community: the agriculture industry. Of those 50 million pounds of antibiotics used in the United States each year, nearly half was consumed by animals. At vast commercial farming operations, tens of thousands of chickens were fed antibiotics in their drinking water if even a few appeared to be sick, a practice that all but assured the spread of resistance as the bacteria of healthy birds became familiar with, and then impervious to, the drugs. Nearly all livestock in America were also fed small, daily doses of antibiotics—"subtherapeutic doses," they were called—as a time-tested, if scientifically unproven, way to make the animals grow faster and fatter. If scientists had tried to devise a means of their own to foster resistance, they could not have come up with a better one than this. The tiny, subtherapeutic doses, also called "growth promoters," enabled bacteria in the animals to get familiar with the drugs but not be remotely threatened by them, and so blithely develop resistance to them. Often, resistance then passed from the livestock to the person who handled the livestock or ate undercooked meat. The agriculture industry had denied this for years, its high-paid lobbyists sounding, as they called for ever more scientific proof, eerily like tobacco lobbyists denying that cigarettes caused cancer.

The most common of the resistant food-borne infections were *Salmonella* and *Campylobacter*. Neither was as virulent as *S. aureus,* the most worrisome bug of all. But both affected so many people that deaths did occur. Each year, *Salmonella* infected 1.4 million Americans and killed 500; *Campylobacter* infected 2.4 million Americans and killed 100. To epidemiologists like Morris looking at the big picture, the more alarming fact was that strains of *Salmonella* and *Campylobacter* were now resistant to as many as five drugs. A relatively new, synthetic class of antibiotics was very effective against both *Salmonella* and *Campylobacter*. Unfortunately, that class was the quinolones, which included drugs being used in livestock. Animal use of the quinolones was provoking resistance in the animals' own *Salmonella* and *Campylobacter*, which were then passing to people who ate that meat. The quinolones included ciprofloxacin, the drug that untold tens of thousands of Americans had persuaded their doctors to prescribe for them as an antidote to anthrax in the aftermath of September 11, 2001. The likelihood of any one of those people receiving an anthrax-laced envelope in the U.S. mail was very, very small. It was extremely likely, however, that many of those people would take Cipro at the first flu or cold symptom they feared *might* be anthrax, accelerating the spread of resistance. An entire class of drugs—the most important new class in four decades— might be compromised far sooner than anyone would have imagined five years before.

The social fabric on which drug-resistant bacteria spread did not flutter to an end at the far edge of town or stop at the city limits. It passed from state to state, country to country, continent to continent. Chaos theory held, famously, that a butterfly flapping its wings in Africa might displace enough molecules around it to set off a series of reactions that resulted in a tornado over Kansas. With drug-resistant bacteria, such a journey was fact: molecular biologists had traced the spread of earlier generations of methicillin-resistant *S. aureus* from a single genetic mutation in Spain, or Australia, or Brazil, clear around the world. Americans and Europeans liked to imagine that they were safe from the myriad infectious diseases that plagued developing nations, and to some extent they were right. Their water was not contaminated by cholera; their air was not abuzz with malarial mosquitoes. But agricultural products carrying resistant *Salmonella* were sent routinely across international borders. For that matter, resistant bacteria could travel across the world in a day by plane, and often did.

Throughout poor and developing countries, the list of other microbes on the march was abysmally long. Either no antibiotics for them were available or, ironically, too many were available, leading to rampant overuse. In China and Mexico, antibiotics were sold over the counter, no prescription needed, like cough drops. More and more, they were about as effective as cough drops, too. New, more powerful antibiotics were needed. But the new drugs—unlike penicillin and its many offspring—were very, very expensive. When two or more had to be combined, the cost rose, usually far beyond what the citizens of poor nations, or their governments, could pay. Treating a single case of multidrug-resistant tuberculosis with a whole coterie of drugs over an infection period as long as twenty-four months cost as much as $180,000 in the United States. In poor nations, the cost might be somewhat less—but so would the levels of sanitation and infection control.

"We are seeing a global resurgence of infectious diseases," U.S. Surgeon General David Satcher warned the U.S. Congress on the eve of the twenty-first century, a dramatic reversal of his office's stance a generation ago. Infectious diseases included viral killers—among them AIDS. But resistant bacterial pathogens were a growing subset, and each threat exacerbated the other. Roughly a third of the world's population, for example, was infected with tuberculosis, the result of early childhood exposure to the bug. Most of those carriers lived their whole lives without having the walled-off tubercles in their lungs break out and cause disease; most remained unaware they even *had* tuberculosis. But as AIDS spread, ravaging the immune systems of everyone it infected, many of its victims then developed active tuberculosis. The more widely TB spread, the more widely, and indiscriminately, a host of drugs were used against it. The more that happened, the more resistant TB became to those drugs.

At the end of a particularly wrenching day on clinical rounds, even Reba McEntire did nothing to soothe Morris as he drove home from the hospital to his tree-lined neighborhood. He would glance at the handsome houses of Roland Park and think, *They have no idea.* Cosseted in their plush living rooms, most of his neighbors simply had no clue how many disease-causing bacteria were growing resistant to antibiotics, how in the silent, invisible war of bugs against drugs, the bugs were beginning to win.

The first thing Morris did when he walked into his big house was go to the kitchen and wash his hands—once more, with soap, just to be sure. Then he went in to hug his wife, a physician herself, and his daughters. Sometimes he marveled at how his daughters took their perfect health—and everything else—for granted. Like most children, they took limited interest in the details of their father's day. Perhaps that was just as well. Morris didn't want to scare them with details of his latest cases. Nor did he want to say that he doubted *their* children would have antibiotics for every need. More and more infections, he felt sure, would be unstoppable killers, just as they had before the age of antibiotics began.

Could it be only a decade ago that most doctors and drug company scientists had believed the antibiotics they had on hand would work forever? In that whisker of time, the entire medical establishment had been forced to swallow a very bitter pill. *No* antibiotic would work forever. Eventually, *every* bacterial pathogen would learn how to become resistant to every drug used against it. Given how quickly bacteria were adapting now, Glenn Morris was only echoing the fears of most colleagues when he predicted that many bacterial pathogens would likely be resistant to all existing antibiotics in another human generation or two. In a decade, after all, while some modest fraction of humanity reproduced itself, bacteria reproduced 50,000 times, trying each time, in some soulless but utterly determined, Darwinian way, to adapt in order to prevail.

How, Morris wondered, could our species have made such a monumental blunder? Sixty years ago, scientists had discovered the first of the natural antibiotics and seen how brilliantly they worked against various bacteria: the biggest medical find of the century. In their excitement, they had failed to remember that bacteria had existed for billions of years, probably before any other life on the planet. Their ancestors, trillions of microbial generations ago, had seen the appearance of brontosauruses, tyrannosauruses, woolly mammoths, and saber-toothed tigers. And they had feasted on their carcasses. After surviving unimaginable extremes of fire and ice, were those bugs really going to let themselves be vanquished by a brand-new arrival in geologic time, using weapons they themselves had devised?

2

IT'S A BUG'S WORLD

As potential victims, we tend, out of fear, to anthropomorphize bacteria. We invest them with malevolent motives and human cunning. We assume they've banded together just to get us, and that all the ills they cause were devised exclusively to torment us into extinction. The foolishness of that conceit comes clear in comparing their residency on the planet to ours. Imagine earth's 4.5-billion-year history as a single year: we appear in the last few seconds of December 31 — perhaps 200,000 years ago. Fungi emerged some 400 million years ago, protozoa about 1.8 billion years ago. They, too, are newcomers compared to bacteria. The oldest known inhabitants of the planet, bacteria have been here at least 3.5 billion years: living, killing, and mutating for nearly all that time without the slightest regard for us.

Archeobacteria, as these prehistoric species are now called, resided first in the water and then in the soil, where many bacteria still live: the seedbed, and life-support system, of so much terrestrial life and biodiversity. There they first developed toxins against the only enemies they had: one another. They fought because they competed for nutrients, light, or oxygen — for ecological niches — and because then, as now, life was about survival of the fittest. Unlucky bugs were zapped out of existence. Lucky ones devised natural shields that were the first resistance mechanisms, precursors of what bacteria do today to defend themselves against human-engineered antibiotics.

From bacteria's perspective, down there in the soil, the arrival of mammals was a threat only in that these lumbering new creatures might swallow them and subject them to killing stomach acids, and almost certainly the bacteria had no fear of that, since they had no fear, or any other emotion, and no consciousness or sense of themselves. To them, mammals were simply a wonderful new source of food and means of survival. But when they hitched a ride and worked their way inside their tasty hosts, they found themselves targeted by some prehistoric, simplistic version of a mammalian immune system. Then they defended themselves as best they could, devising new variants of the chemical weapons and shields they'd used against other bacteria in the soil. Ironically, those with the most overwhelming virulence factors often fared the worst: they killed their host so quickly that they didn't have time to jump to the next inviting nutrient source, and so they died aborning. Many of the more successful ones fought to a draw with their host's immune system, and over millions of years joined the pack of friendlier neighbors along for the ride. To use the microbiologists' term, they became human commensals: bacteria that reside on us, or in us, and often help us survive.

Our skin, in fact, is covered with bacteria—about a trillion of them—which form a layer that protects us from other, less benign microbes. Tens of trillions more reside in our alimentary canal, from our stomach down through our intestines to our rectum; there they help digest food and release nutrients that our bodies can absorb from our intestines. Bacteria also gather in our mouth and nose and eyes, drawn by the nutrient-rich warmth and wetness; in and around our genitalia; in the webbing between our toes. In all, they comprise about one twentieth of our body weight. That means President George W. Bush is carrying around about nine pounds of bacteria, actress Julia Roberts about six pounds, basketball player Shaquille O'Neal perhaps twelve pounds.

Bacteria also reside on every surface of our homes. Though most cause no ill effects, a recent vogue for "antibacterial" household cleansers has swept the country. The cleansers, which now predominate in the household sections of every supermarket in America, make the pitch that bad bacteria can be wiped away from kitchen counters and bathroom sinks like so much invisible dust. To any microbiologist, the pitch is absurd. Even if the "antibacterial" cleansers do eliminate certain bacteria that other cleansers do not—a difficult claim to prove—the cleared field is immediately filled by *other* bacteria, some of which may be more harmful than those the cleanser wiped away. And as Tufts's Stuart Levy has observed, all bacteria targeted by "antibacterial" cleansers will grow resistant to those cleansers eventually, becoming hardier pathogens as they do.

Bacteria are, in fact, among the most abundant organisms on earth. Run from your bacteria-filled house with your "antibacterial" cleansers in hand, and you'll step on soil inundated with them to uncountable orders of magnitude. A single pinch of dirt holds more than a billion of them. Some soil bacteria convert nitrogen into essential nutrients for plants. Others decompose dead organisms. Some bacteria migrate through the air, others in the water. Here, too, they break down dead matter. Without bacteria, all the human corpses in history would still be with us. So would the garbage and wastes of every person who ever lived. Without bacteria, the seas would be filled with dead fish and their wastes.

Bacteria accomplish all this because, like well-equipped sailors, they carry every tool they need within their one-celled vessels. Their cell walls encase cytoplasm, a thick brew of proteins and other essential ingredients. Floating within that cytoplasm is the bacterium's chromosome, a closed DNA loop of genetic information containing the bacterium's genes.[1] The genes contain all the traits that define a bacterium as the kind that it is, as well as the directions that tell it how to do what it must to survive and replicate. All these orders are carried out by proteins that the DNA's genes "encode." When a pro-

---

[1]In some bacteria, such as the Lyme disease spirochete, the DNA is linear rather than a closed loop.

tein is needed, the DNA instructions for it are copied by the ever helpful middleman, RNA. The RNA, in turn, conveys these orders (on an RNA molecule with essentially the same sequence as the DNA gene) to little protein-producing "factories" in the cytoplasm, called ribosomes.

A bacterial cell's wall is its shield and strength, protecting it against other bacteria and, when it bobs around in animals or humans or the environment, from osmotic pressure that might obliterate it. It also fundamentally distinguishes bacterial cells from animal and human cells, which lack such walls. That distinction makes the cell walls of bacteria excellent targets for antibiotics. A drug can be designed to attach to one part of the wall or another, inhibiting the bacterium when it tries to replicate and build more wall but doing no such damage to the human cells surrounding it. A drug that keeps the bacteria from replicating is called bacteriostatic; a drug that kills the bacterium outright, often by causing its cell wall to break, or "lyse," is called bacteriocidal. The reason antibiotics have no effect on viruses is that viruses lack walls and all the other cell-building targets that bacteria have. When doctors prescribe antibiotics for viral infections, they might as well be prescribing M&Ms.

Bacteria have the same basic needs we do: to eat, excrete,[2] protect themselves from harm, and, with just as much urgency as we feel but with hardly any of the fun, reproduce. Small as they are, undistracted by the yearnings for love, family, houses, cars, new clothes, and the occasional cocktail, they manage those tasks rather more quickly and efficiently than we do. In ten hours, a single bacterial cell can produce more than a billion daughters, as scientists call the offspring; after four more hours, that one bacterium will have undergone enough divisions to equal the entire human population of the earth. In the almost twenty years since a new class of antibiotics appeared, most bacteria have undergone about 100,000 generations. Under the barrage of antibiotics, besieged bacteria need only produce a few random mutants capable of resisting a particular drug for the bug to begin to

[2]Technically, they *secrete* various enzymes.

fight back. The mutants proliferate, filling the niche left clear by the drug's effects on the mutants' susceptible brethren, and—presto!—a resistant bug takes hold.

Most bacteria take one of three general shapes: the rod, the sphere, and the spiral. All are fully capable of finding ways to cause us harm, either in the bloodstream or on the skin. Some bacteria form hard little capsules called spores that break off on their own when the bacteria are threatened and wait, in a dormant state, until conditions improve. Spores have remarkable powers, both of resistance and pathogenicity. They can survive two hours in boiling water, when all other bacteria die. They can survive in a barely more hospitable environment for twenty years or more, then reactivate. Four of the deadliest diseases known to man are caused by bacteria that produce spores: botulism, gas gangrene, tetanus, and anthrax.

Optimists reason that ultimately even the bacteria most pathogenic to us will grow benign and become commensal with us because it's in their interest to do so: a live host is better than a dead one. Pessimists observe that that might have been true before humans harnessed antibiotics for their own use—an incredible onslaught, the bacterial equivalent of a nuclear war—but since then, bacteria have fought back fiercely, growing more resistant as a result. Many of the bacteria that threaten us today were actually benign before the golden age of antibiotics; others, like enterococci, were simply irrelevant because more virulent bacteria claimed all the human ecological niches. (As early antibiotics knocked out the more virulent bugs, the field was left clear for tamer ones like enterococci to spread.) And nearly all those that threatened us then have learned to be multidrug resistant now, so that their threat is more potent. Of those, the most dangerous is the one that has succeeded in becoming resistant faster than any of its fellow species to one antibiotic after another: one of man's oldest and most enduring enemies, *Staphylococcus aureus*.

More than a third of us are carrying *S. aureus* right now, either on our skin or, most likely, in our nostrils, because that area is moist, somewhat sheltered, and supplied with water and nutrients the bac-

teria need. The rest of us have all but certainly carried it in the past
and will carry it in the future, along with a dozen other species of
Staphylococcus. Fortunately, those other species are basically benign, and most of us will merely be *colonized* by *S. aureus,* not infected by it. But in a host with an impaired immune system, *S. aureus* is an enemy standing by, waiting for a first cut in the skin or a
chance to invade the respiratory tract, to dive into the tissues of the
weakened host and start spreading serious, potentially life-threatening infections.

The power of *S. aureus* derives from its array of what scientists
call virulence factors. Other bacteria have them, too, but usually to
a lesser degree. *S. aureus* comes in like a SWAT team armed to the
teeth, with every one of its fine-honed weapons primed for immediate action. It has proteins that act as advance guards to infect tissues
around its landing site. As it slips inside the skin, it has toxins that
tear through surrounding cell membranes like so many machine-
gun bullets. When the immune system responds by dispatching
mittenlike phagocytes, wide-ranging cells that surround an invader,
S. aureus has a way of coagulating blood around itself as protection.
The toxins keep coming: one kind to blast the immune system's red
blood cells, another to help it evade various white blood cells. Perhaps
the bug's most arcane virulence factor is a special protein that binds
backward to immune system antibodies—meaning that now the
bacteria are wearing the antibodies as a disguise, like the elephants
in *Babar* that paint eyelike rings on their buttocks and edge backward over a hilltop to appear as monsters to the enemy rhinos below. As it fires all these weapons and plays all these tricks, *S. aureus*
keeps multiplying. The more progeny it produces, the sicker it
makes its host.

The immune system of an otherwise healthy victim tends to overwhelm *S. aureus* in time to limit the damage to skin pustules, boils,
carbuncles, and impetigo. But in an immunocompromised patient, *S.
aureus* causes wound infections of various kinds, pneumonia, heart
infections, and meningitis. If it spreads into the bloodstream, it can
be rapidly lethal, causing septicemia—infection of the blood—that

knocks out internal organs and may produce septic shock. Newborns are vulnerable to "scalded skin syndrome," in which *S. aureus* toxins cause exfoliation that painfully exposes sensitive red underlayers of skin to air and other infections. However they choose to kill, *S. aureus* bacteria are so virulent that very few are needed to do the job. (Enterococci, by way of comparison, need 10,000 times more bacteria to kill someone.) With its arsenal of virulence factors, *S. aureus* produces a wider range of infections in a broader cross section of people than any other bacteria. Ecologically, it's the most successful of all bacterial pathogens and the number-one cause of hospital infections in the world. Each year, it affects as many as 9 million Americans to various degrees, according to the CDC.

More likely than not, *S. aureus* evolved from rodlike soil bacilli, the earliest known bacteria. At some point the plucky bugs clambered aboard primitive vertebrates as they emerged from water to crawl through the dirt. Eventually, *S. aureus* seems to have found a more comfortable home on the skin of certain species of monkeys. Today it lives among, and preys on, various orders of mammals, from rodents and rabbits and birds to cats and dogs, sheep and horses and cattle. With its particular virulence factors, it's exquisitely engineered to combat the human immune system and dine on human nutrients. Yet the bacteria are not exclusively mammalian parasites. They can live in water, on decaying matter, on almost any surface. Unlike most bacteria, which quickly dry out and die, *S. aureus* can also endure in adverse circumstances, without any food source, for years at a time. Researchers have been known to take an agar plate with *S. aureus* on it from storage after a decade or more and find that with a drop of sheep's blood or some other nutrient to revive it, the *S. aureus*'s spherical clusters glow bright yellow—and then start growing again. *S. aureus* can endure because it has incredibly thick cell walls, and the highest internal pressures of any bacteria, which helps it withstand the external pressure exerted by the bloodstream. And, a microbiologist might be tempted to say after a long night in the lab, because it seems so *determined*.

S. aureus is a master of disguise, hiding behind a multitude of disease symptoms, and so the earliest recorded evidence of it is less clear than that of, say, smallpox or tuberculosis. In the Bible's book of Exodus, a suspiciously *Staph*-like condition is mentioned: "And they took ashes of the furnace, and stood before Pharaoh; and Moses sprinkled it up toward heaven; and it became a boil breaking forth with blains upon man, and upon beast." (Blains are swellings or inflammations; boils are typical in staph infections; and so blains may have indicated the harsh symptoms of *S. aureus*.) The Romans referred to "good and laudable" pus ("*pus bonum et laudabile*"), which was probably a reaction by the white blood cells to *S. aureus* wound infections. But the very notion of germs—their unseen existence, their role as disease carriers—did not emerge until the mid-sixteenth century, when Italian physician Girolamo Fracastoro propounded it in the second volume of a treatise entitled "Contagious Diseases and Their Treatment." Fracastoro described furuncles, boils, and various other infections; probably the germ causing such ills was *S. aureus,* though he could not see it. Not until 1683 did Anton van Leeuwenhoek actually observe through his handmade microscopes the spherical "animalcules" that were likely some species of *Staphylococcus.*

Two more centuries of undiagnosed infection and disease passed—of boils, bloodstream infections, and suppurating wounds that usually went untreated, save by Chinese and African folk doctors who reportedly used moldy soya beans and bread as the first antibiotics—before Alexander Ogston, a Scottish physician, observed clusters of spherical bacteria closely enough through his own microscope to christen them *Staphylococcus.* (*Staphyle* is Greek for "bunch of grapes," which is what they look like under the microscope, because they form clusters; *cocci* means "spherical bacteria.") Ogston went on to show that some kinds of *Staphylococcus* had yellowish colonies: these he called the golden *Staphylococcus—Staphylococcus aureus*. He distinguished these cocci from other bacterial spheres that were arranged in chains: those he called *Streptococcus.*

In France in that very year—1880—Louis Pasteur was making the same observations, though letting his "germ theory" carry him to other experiments, from saving the wines of France to discovering a vaccine for rabies. But if he and his brilliant German peer, Robert Koch, who identified the tubercle bacillus in 1882, did little directly to further the investigation of *S. aureus* in particular, they helped in a general way by ushering in the golden age of microbiology: the study of microorganisms.

One other contemporary of Pasteur's helped simply by organizing bacteria into two groups. Christian Gram, a Danish doctor, declared in 1884 that when he stained bacteria with a certain solution that turned them blue-purple, then tried to remove the dye, some bacteria turned transparent again and others retained the dye. Those that turned transparent were eventually called "Gram negative," and those that remained blue-purple were "Gram positive." As it turned out, Gram positives have single-layer cell walls, among other traits, and include *Staphylococcus, Streptococcus*, and *Enterococcus*. Gram negatives don't have cell walls but do have two-layer cell membranes; among them are *Pseudomonas aeruginosa, Klebsiella pneumoniae, Acinetobacter baumannii*, and *Escherichia coli*. The nomenclature sounds pedantic, but it became a shorthand for the traits associated with one group or the other, including their susceptibility to particular antibiotics. In microbiology, it has become as much a defining aspect of bacteria as *Mr.* and *Mrs.* are to the identification of people.

In the years leading up to and following World War I, *S. aureus* took lives around the world—more lives, arguably, than pneumonia and tuberculosis, because behind its many masks it caused an awesome range of illnesses. The global flu epidemic of 1918 that took 20 million lives had more obvious causes—chief among them the dreadful conditions for wartime patients—but *S. aureus* played a contributing role. With their systems weakened by viral infection, many of the sick were likely killed by *S. aureus,* which worked its way from the lungs into the bloodstream. And soldiers with open wounds were open season for *S. aureus* through cuts in the skin.

It was in the aftermath of that terrible scourge that British scientist Alexander Fleming discovered, by chance, a natural antidote to certain bacteria. No antibiotics existed yet: the word, which means "against life" because bacteria use their toxins to kill one another, had not been coined. But Fleming had observed that something in human tears caused some bacteria to lyse. In itself, this was a finding of little use, because "lysozyme," as Fleming called it, seemed to work only against nonpathogenic bacteria. And in the 1920s, Fleming had no means to isolate what "lysozyme" was, much less produce commercial quantities from human tears. But it reaffirmed a fascinating concept: that humans, animals, and probably plants and microbes might contain antibacterial substances—indeed, they must, Fleming reasoned, or pathogenic bacteria would have wiped them out eons ago.

So famous is the story of Fleming's subsequent discovery of penicillin on a September morning in 1928 that its details vary from telling to telling, as in an Arthurian legend, provoking some lingering uncertainty as to how much is true. Did the scientist really leave his window open at St. Mary's Hospital when he went on vacation? And did he return to notice that something appeared to have floated in through the window from Praed Street onto one of the round agar plates on his lab bench? In *The Birth of Penicillin,* Ronald Hare concludes that the window was probably *not* left open, first because a bacteriologist devoted to careful work amid sterile conditions would hardly be apt to expose his specimens to the outside elements, and second because, as Hare determined after admirable sleuthing, that particular window was hard to open. But unquestionably, one of the agar plates on Fleming's bench contained *S. aureus,* and, unquestionably, an airborne spore of the common mold *Penicillium* landed on it, lysing the nearest bacterial cells to create a "zone of inhibition."

Chance favors the prepared mind, Pasteur had declared, and so Fleming was favored. But after publishing an observation about *Penicillium*'s effect on his plate of *S. aureus,* he seemed to let the matter rest. Privately, he did make efforts to test the fungus on infected patients, but as a pure researcher he had trouble marshaling supplies of *Penicillium* for use as needed, and he was unable to purify his "mould juice." One of Fleming's former students, Cecil George Paine, became the first to achieve a cure with the fungus, using it in

1930 on a patient's eye infection, but as a clinical doctor Paine too failed to follow up on its potential.

More than a decade passed before two London-based researchers turned *Penicillium* into penicillin. Howard Florey, an Australia-born medical researcher working at Oxford, and Ernst Chain, a German Jew who had emigrated to England to escape Nazi oppression, began by successfully re-creating Fleming's experiments with lysozyme. They managed partial purification of a *Penicillium* sample, and on May 25, 1940, injected it into four of eight mice after first injecting all of the mice with a virulent strain of *Streptococci*. The mice untreated with *Penicillium* died; the inoculated mice survived. Elated, the researchers turned Florey's laboratory into a crude factory, employing "penicillin girls" to do the tedious work of manually purifying "mould juice" into penicillin. Their first human test case was an English policeman suffering from a severe bacterial infection. On December 27, 1940, a precious sample of penicillin, still only partially purified, was injected into him. The policeman's condition improved until the researchers' supply of penicillin ran out; then the infection spread, and the policeman eventually died. But in several other cases that winter and spring of 1941, penicillin saved lives.

Amazingly, the new drug killed all the classic battlefield infections— gangrene, septicemia, and pneumonia—as well as a wide range of other Gram-positive infections: *Staphylococcus,* streptococci, tetanus-producing *Clostridium,* and the syphilis spirochete. Moreover, it had no toxic side effects other than an allergic reaction in rare cases. Its only shortcoming was that it had virtually no effect on Gram-negative bacteria, among them *Klebsiella pneumoniae* and *Pseudomonas aeruginosa.* In the thick of war, with Nazi bombs raining down on London, the British pharmaceutical industry was unable to scale up production of the drug. But a huge American effort was soon under way, and by 1944 penicillin was saving so many soldiers' lives that, eventually, the war would be said to have been won, in part, because of it.

With the success of penicillin, the race for antibiotics was on, a race engaged by highly educated scientists sifting around in dirt and, occasionally, sewage water. The first of the cephalosporins, today the most widely used drugs in the world for a huge variety of illnesses, including pneumonia, meningitis, and middle-ear infections, came from sewage water off Sardinia in 1945; human waste is thick with bacteria and therefore, scientists rightly reasoned, likely to contain natural antibiotics produced by various gastrointestinal bacteria to zap one another. Chloramphenicol, used for typhoid fever, emerged from soil in Venezuela in 1947. Erythromycin, good against pneumonia, came from soil in the Philippines in 1952.

Amid the excitement came the first warnings of resistance to antibiotics. Ernst Chain and a colleague named Edward Abraham issued the first report even before penicillin reached the market. They had found that when certain strains of *S. aureus* were exposed at length to penicillin, they produced an enzyme that destroyed the penicillin before it could bind to the bacterium's cell wall and disrupt cell replication. Fleming himself did similar experiments and declared publicly in 1945, only two years after the drug had reached the market, that indiscriminate use of penicillin would lead to widespread resistance among all Gram-positive bacteria. "The greatest possibility of evil in self-medication," he told the *New York Times,* "is the use of too small doses so that instead of clearing up infection, the microbes are educated to resist penicillin and a host of penicillin-fast organisms is bred out which can be passed to other individuals and from them to others until they reach someone who gets a septicemia or a pneumonia which penicillin cannot save."

But no one seemed to hear. Not doctors, not patients, and certainly not the burgeoning drug industry. Penicillin was recommended for every imaginable ill as an over-the-counter remedy without prescription until the mid-1950s. It was put in cough lozenges, throat sprays, mouthwashes, and soaps, all available over the counter. It was even put in drinks as a general health powder. Penicillin's elevation to The Great Panacea was, to be sure, understandable. The drug did do wonders, and nowhere more so than in the operating room, where an injection of penicillin before and after surgery reduced mortality rates dramatically—enough that a whole new realm of complexity in medical procedures suddenly became possible. Overall, Americans whose lives had already been lengthened by improved sanitation and other health measures gained an average decade or more from antibiotics and the decline of various infectious diseases: from about fifty years in the nineteenth century to about sixty-five years in the mid-twentieth century. Yet the warnings were proving prophetic. By 1946, 14 percent of the *S. aureus* strains isolated in U.S. hospitals were penicillin resistant. A year later, that figure rose to 38 percent. One year after that, it was 59 percent.

Exactly how penicillin worked and how the bugs resisted it took years to decipher. Penicillin, it turned out, mimicked a particular element in the cell wall–building process in Gram-positive bacteria, a structure of five amino acids called a peptide, which acted as a cross-link in the burgeoning wall. The bacterium had construction workers, in effect, to put those cross-links in place; the workers came to be known as penicillin-binding proteins, or PBPs. The penicillin acted as a decoy peptide, and when the workers tried to use the drug as a cross-link, they failed. Without cross-links, the cell's new wall ruptured, or lysed.[3] Various strains of *S. aureus* and other Gram-positive bacteria managed to trump the drug, and others from the beta-lactam class of antibiotics, by producing an enzyme eventually called beta-lactamase. The enzyme sundered, or cleaved, penicillin's lactam ring structure as neatly as a knife cuts a wedding cake.[4]

Beta-lactamase enzymes, like any other part of the cell, had to be encoded by genes. For scientists, finding the first resistance genes for beta-lactamase enzymes was disconcerting: it suggested that the bugs had either reactivated an ancient gene used to combat the *Penicillium*

[3]Strictly speaking, that was the first step of a two-step process. When the penicillin kept the bacterium from forming new cell wall, it also triggered the release of other enzymes in the bacterium meant to disrupt the old cell wall. These autolysins, as they were called, made the bacterium lyse.

[4]Bacterial enzymes that stymie antibiotics are usually given the name of the chemical they confront, plus the letters "ase," so the beta-lactam ring of penicillin was sundered by beta-lactamase.

fungus in nature or managed a successful random mutation, only a short time after being exposed to penicillin, that conferred resistance to the drug. But even a recent mutation could be shrugged off, scientists felt, given how rare such mutations were. Indeed, the odds of a resistance gene occurring by mutation were pegged at once in every 10 million to 100 million multiplications of bacteria. If a second drug was used in combination with the first, the chances of a bug developing a spontaneous mutation to *both* drugs were truly infinitesimal.

Then in 1952 came a startling case in Japan. Researcher Tsutomu Watanabe reported that he had isolated a strain of *Shigella,* the dysentery bacillus, that was resistant to several antibiotics at the same time! Chloramphenicol, tetracycline, streptomycin, and the sulfonamides—it thwarted them all. The odds simply did not allow multidrug resistance by the spontaneous mutation of chromosomal genes. Something else had to be happening—and, indeed, something was.

This more complex scenario, it turned out, began with little loops of DNA called plasmids that floated in bacterial cytoplasm like minichromosomes, carrying additional genes. Some of these genes were helpful in a general way to many bacteria: one enabled bacteria in the gastrointestinal tract to attach themselves to the cells of the tract's lining to keep from being swept away by the ongoing flow of digested food. Others turned out to carry "R factors," genes for antibiotic resistance. The Japanese researchers began working on a hunch that these genes could be transferred from plasmids to a cell's chromosome like so many spare parts, or even from one bacterial cell to another. Eventually it was discovered that genes on plasmids could indeed move to different places on the chromosome as needed. Gene-bearing plasmids could also jump to another bacterial cell altogether when the two cells adjoined in what scientists called conjugation—bacterial sex. Even smaller lengths of DNA called transposons—like bits of free-floating string—could ride into a new cell on a plasmid. The plasmid itself might not survive if the cell was of a different species, but the transposon could latch on to the new cell's chromosome, bringing its resistance genes with it. For that matter, transposons could even ride in on little viruses called phages, a process called transduction. And if these tricks failed, a cell could release DNA into the environment and have another cell simply absorb it, as if one had passed a football to the other, a process called transformation. However the passage was accomplished—conjugation, transduction, or transformation—bacteria began to gather genes of resistance to different antibiotics from all over the bacterial kingdom, as if collecting charms on a charm bracelet.

As the 1950s unfolded, *S. aureus* used these different tricks to resist every new drug thrown against it: streptomycin, chloramphenicol, bacitracin, neomycin, aureomycin, erythromycin, and others. Using two drugs together seemed to deter the bug at first—penicillin and streptomycin was one such promising combination—but soon *S. aureus* developed resistance to both. Some strains became not only resistant to streptomycin but dependent on it. Meanwhile, resistance to penicillin and other, newer beta-lactams like ampicillin kept growing, especially when the drugs began being given orally and their use became as widespread as that of chewing gum. By the mid-1950s, roughly half of all *S. aureus* infections in hospitals were resistant to most available antibiotics. And for those people unlucky enough to contract them, the mortality rate from systemic *S. aureus* infections— the invasive bloodstream infections—was about 60 percent.

The answer, it seemed, was to develop semisynthetic—human-engineered—chemicals that would be entirely new to *S. aureus* and other bugs. There would be no resistance genes for those, either on a bacterium's own chromosome and plasmids, or from somewhere else. The three most promising semisynthetic drugs, nafcillin, oxacillin, and methicillin, all had the extra advantage of being unaffected by penicillin's beta-lactamase enzymes because their beta-lactam rings could not be sundered by them. Scientists had designed them that way.

Methicillin, issued in the early 1960s, was among the first of these beta-lactamase-resistant penicillins—a seemingly devastating weapon against most Gram-positive bacteria.[5] Yet just one year later, the first case of methicillin-resistant *S. aureus* (MRSA) was reported. This

---

[5] The enterococci were one notable exception.

time, because the drug's chemical structure was modified by humans, there could be no prehistoric gene of resistance that *S. aureus* found on its chromosome or imported from another common bug. Methicillin resistance was a brand-new response to a new, semisynthetic substance. That was shocking.

Years would pass before this revolutionary resistance mechanism was fully understood. At Cambridge University in England, a biochemist named Peter Reynolds would determine that *S. aureus* had brought in a new resistance gene, the *mecA* gene, on a transposon, apparently from a distant relative of *Staph* found in squirrels. The *mecA* gene encoded a new cell wall protein that managed *not* to bind to methicillin and thus built strong cell wall cross-links despite the drug's presence. In so doing, *mecA* conferred resistance not just to methicillin but to every other beta-lactam drug, from penicillin to the cephalosporins. Spreading on its own from ward to ward of hospitals all over the country as the 1960s unfolded, MRSA caused painful infections and killed with impunity. For this, humans could thank themselves. MRSA was entirely their creation. It wouldn't exist without the drug that spawned it, since without a need to defend itself against methicillin, *S. aureus* wouldn't have retained the *mecA* gene that it had imported as a random mutation.

Ever since the discovery of streptomycin in 1944 from a genus of soil bacteria called actinomycetes, drug companies had sifted the soil, like so many gold miners, in search of the next great natural antibiotic. Who knew how many kinds of bacteria might live in the soil? A million was one educated guess. Yet of those, only a few thousand had been identified as yet, and elements of just a handful turned into drugs. How many of those bacteria might have antibiotics, developed over billions of years for use against their microbial enemies, that could knock out *S. aureus, S. pneumo,* and the worst of the other bacterial infections spreading so worrisomely through hospitals around the world? Drug company researchers studied dirt from their backyards. They asked friends to scoop up dirt on their travels. And they looked for volunteers in far-off places willing to send in samples on a regular basis. For that, decided Edmund Kornfeld, a chemist at the Eli Lilly company, missionaries were especially good prospects.

Through the Christian and Missionary Alliance of New York City, Kornfeld wrote to the Reverend William W. Conley, a former army chaplain who had settled with his wife in Borneo as part of an Indonesian mission. He explained that in 1951 Lilly had begun a botanical screening effort to sift for natural antibiotics and that he needed the help of expatriates like Conley to gather soil samples. Kornfeld sent along several sterile vials and explained how to label them. Within a month, Conley sent the vials back full, and wrote, "Certainly hope there is something promising in our dirt out here. From our past attempts at gardens, I cannot say it has been promising."

From that first batch, Kornfeld isolated an antibiotic-producing organism from the *Streptomyces* species of actinomycetes from which the drug chloromycetin could be produced—a drug already discovered, unfortunately, but a promising finding nonetheless. Kornfeld wrote back with more empty vials and asked Conley to collect samples "off the beaten track . . . away from the towns and villages." Conley left on a furlough before he could fill the new vials, but he delegated the task to the Reverend William M. Bouw. In February 1953, Bouw dutifully sent this second batch of vials back, filled with dirt from a variety of sites in Borneo. One contained an intriguing new member of the *Streptomyces* species. The Lilly team christened it *Streptomyces orientalis* and began to experiment with it on various bacteria.

On June 18, 1953, Lilly researchers marked one test tube M4 3–05865 to indicate that it was the number 5,865 isolate subjected to the test tube method. When a biochemical "fingerprint" was made of the sample, showing its efficacy against different standards, the researchers were astounded. Number M4 3–05865 had more antibiotic activity than any sample they'd seen.

For the next several months, the researchers tried to learn what nutrients this new organism needed. That was critical: if the organism could not be sustained, it could hardly be developed for large-scale production as a drug. The new bug turned out to be finicky in its tastes but did enjoy milk derivatives, various sugars—and Brer

Rabbit brand molasses. As important, the researchers determined that the amount of Number M4 3–05865 needed to make it bacteriocidal, capable of killing, was only a bit more than the amount needed to make it inhibitory, or bacteriostatic. With many antibiotics, that wasn't the case. If only a small portion was needed to keep the bacteria from growing but much more was needed to kill it, then the bacteria would have time to get familiar with the new threat and possibly grow resistant to it. With Number M4 3–05865, the bacteria would get the equivalent of a one-two punch. They'd simply have no time to develop resistance before more Number M4 3–05865 knocked them out.

But as Florey and Chain had learned with penicillin, ratcheting up the drug from in vitro to in vivo—from test tube to lab animal and human use—was harder than just whipping up a bigger batch of bug stuff. Not all the antibiotic samples gleaned from 05865 were exactly the same. Some were more potent than others. And some were purer than others. Penicillin, at least, had dissolved readily in water, which meant it could be administered by injection without difficulty. But the antibiotic from 05865 was a large molecule, not very soluble at all. That meant that it could not pass through the gastrointestinal tract into the bloodstream and so could not be taken orally. And when the researchers finally managed to purify enough of it to use on two human patients, it caused pain on injection. The patients' infections did improve, however, and with more tests the researchers began to see that 05865 might have a range of efficacy as broad as penicillin's.

The next step was to give samples of what they were calling vancomycin (later, no one would quite remember who had come up with the name, from the Latin verb *vanesco,* to vanquish) to a few doctors around the United States for them to use with chronically ill patients not responding to other available drugs. Vancomycin worked fairly well, the doctors reported, though not brilliantly. Against *S. aureus* it was a reliable but not very fast killer—not nearly as fast as penicillin had been against susceptible strains. Also, because of that large molecular size, it failed to reach all tissues equally well. It was almost useless, for example, against meningitis. And it had serious side effects:

not just pain around the area of injection but phlebitis, an apparent risk of damage to the kidneys as the drug was excreted, and, in a few cases, partial deafness.

Just as the team was isolating the purest, most potent, and least toxic samples of 05865, new soil from India came in with better vancomycin cultures. After much debate, three years of research on 05865 were junked and the same painstaking tests begun with the new samples. As with the first batch, an elaborate eleven-step process of purification was needed to produce a chemically consistent antibiotic. Still the yield contained a mix of fractions, as the technical term had it—not good—and still there were side effects. Because of its low solubility, and to minimize the pain associated with administering it, vancomycin could be used only in hospitals via slow, intravenous drips into a major vein, and it had to be administered four times a day. But the side effects were sufficiently confined and short-lived that the Food and Drug Administration approved it anyway in November 1958.

Vancomycin reached the market with a dubious reputation, tainted by its toxicity in clinical tests. Those doctors not put off by its side effects balked at the drug's price: at roughly $100 a day per patient when it first appeared, it was exponentially more expensive than penicillin and the other beta-lactams, which cost literally pennies a day. When methicillin was licensed two years later, in 1960, vancomycin was all but forgotten.

Methicillin resistance, creeping up through the 1960s in Europe, not only shocked doctors but scared them. It made them wonder, for the first time, if a magic bullet simply might not exist for bacteria so nimble and resourceful. Perhaps, no matter what was thrown at them, the bugs would always find a way to resist it. The doctors also faced an immediate problem. With methicillin-resistant *S. aureus,* or MRSA, the *mecA* gene passed from strain to strain, and species to species, conferring multidrug resistance as it went. By the early 1980s, as MRSA became a widespread pathogen in U.S. hospitals, vancomycin's toxicity no longer seemed so significant. And its appeal was dramatic: it was the only drug that worked consistently against MRSA.

Meanwhile, Lilly's researchers had been working on the drug's poor solubility. They found that with hydrochloride salt, the drug dissolved more easily in water. They saw, too, that if the drug was just administered in lower doses, its side effects decreased substantially. Those side effects would still make it a drug of last resort, but when it did get pressed into use, it knocked out every Gram-positive bacterial threat it came up against. Among its most dramatic successes were cases of bacterial endocarditis—heart infection—usually caused by *S. aureus* or *Streptococcus viridans*. Before vancomycin, acute methicillin-resistant staphylococcal endocarditis was 100 percent fatal. Vancomycin knocked out virtually every bacterial heart infection it was used against.

To everyone's relief, vancomycin was a drug on the shelf just waiting to be taken down: a prospect that needed only slight tinkering to make it serviceable, without a years-long process of clinical testing ahead of it. And neither *S. aureus* nor other bacteria appeared remotely able to develop resistance to it. Like penicillin, the drug inhibited cell wall synthesis. Unlike penicillin, it inhibited a whole structure of cell wall cross-linking, in three dimensions, like a big mitt. To become resistant to vancomycin, the bacteria would have to do more than bring in a single resistance gene on a plasmid or use a single, random mutation of their chromosomes. They would need to change each of several enzymes in the cell wall to which vancomycin attached, then build a *new* cell wall cross-link structure with new enzymes, presumably by importing a whole new set of genes to encode those enzymes *and* figuring out how to organize those genes in such a way to get the job done.

The odds against any bacterium being able to effect all those changes and have them work in concert were astronomical. By any reasonable standard of mathematical logic, it simply wouldn't ever occur.

3

EARLY WARNING

Glenn Morris had the makings of an epidemic on his hands, right here in his own medical center in downtown Baltimore—an epidemic of once-humble enterococci, resistant now to every antibiotic in the hospital formulary, including vancomycin. An epidemic that was starting to kill.

It had started so subtly, like the first gentle offshore breezes that only longtime islanders can recognize as the harbingers of a hurricane. Back in 1989, the lab for the affiliated University of Maryland and Veterans Affairs hospitals had reported what appeared to be one of the first cases in the United States of vancomycin-resistant enterococci, or VRE. The infection had afflicted a very sick, elderly patient who had undergone long courses of vancomycin for various ills, and, frankly, Morris was unsurprised. He had always scoffed at the cult of vancomycin as a miracle drug to which resistance would never arise, and he imagined that an immunocompromised patient treated for weeks with the drug was exactly the sort of candidate in which resistant enterococci would breed. For some time after that, the cases of VRE trickled in sporadically enough to seem insignificant. Nearly all the deaths, as it happened, occurred in the oncology ward. But if the actual numbers remained low, the percentage of increase began to seem ominous. As the incidence of mortality increased as well, Morris began to see that the scope of the problem was much, much larger than he'd ever imagined.

By 1993, VRE had become epidemic in all Baltimore hospitals and was actually growing endemic. Epidemic meant one, or perhaps a few, types or clones of VRE replicating at such a rapid pace that the bug affected far more patients than anyone had expected it would. Endemic meant that the epidemic wasn't going away—it would be present in the community from now on—and it no longer consisted of one or a few clones. The bug had settled into its human hosts long enough and successfully enough to pass its resistance genes on to various other enterococci swimming around in the hosts' gastrointestinal tracts so that the hospital was now beset by scores of strains. The CDC reported that VRE in U.S. hospitals had shot up from 0.3 percent of all enterococci in 1989 to 7.9 percent in 1993. In some hospitals, like Morris's, the percentage was closer to 14 percent. Spanning a century, this increase would have been notable. In four years, it was extraordinary.

Once again, the bugs had beaten the odds. VRE was still a second-rate pathogen, but, as one doctor said, it was now a first-rate problem because it was all but untreatable: strains of *Enterococcus faecium* were completely impervious to any and all antibiotics.[1] As an enteric bug— one living in the gut—it could easily slip out to cause life-threatening peritonitis: inflammation of the membrane lining the abdominal cavity and covering various organs. It also caused abdominal abcesses and urinary tract infections; it seeped into the blood to cause bacteremia and heart infections (endocarditis). By the mid-1990s, the mortality rate for patients with VRE in Morris's hospitals had climbed to more than 40 percent. Some of these might have been dying of other conditions: dying, therefore, *with* VRE, not *because of* it. But VRE appeared to be a determining factor in many if not most of their deaths.

---

[1] Some strains of VRE remained susceptible in vitro to certain antibiotics, particularly chloramphenicol and doxycycline. But the efficacy of these drugs against VRE in patients remained unclear; basically, the bug was still unstoppable.

Whether it infected or merely colonized its host, VRE, like many bacteria, sent some of its troops out on a daily basis in what doctors delicately termed "fecal carriage." The bugs were amazingly hardy in the outside air and could pass easily from feces to a healthcare worker's hands to the next human host. This was a danger considerably heightened by diarrhea, which came as an almost inevitable side effect of heavy antibiotic use. (The diarrhea, in turn, flushed vital nutrients from an already weakened patient. Sometimes a patient literally died as a result of antibiotics.) Unlike methicillin-resistant *S. aureus,* the VRE strains could even survive for lengthy periods on medical instruments, bedrails, on gloved or ungloved hands, on telephones and countertops. A particularly insidious route for VRE was in the biofilm, as it was called, that coated intravenous and urinary catheters, prostheses, and endotracheal or gastric tubes. VRE was like the oobleck in Dr. Seuss's classic children's story: gunk getting over everything, except that it was invisible, and, unlike oobleck, it killed.

Over a two-year period, Morris conducted one of the country's first large studies of VRE in what were, in fact, two of the first hospitals in the United States to be hit by the epidemic. The number of active VRE infections he recorded seemed modest—75 out of nearly 60,000 patients admitted to the twin hospitals—but as an increase from nearly zero four years before, it was alarming, especially given that it appeared that nearly half those patients died of their infections. The greater shock was that on three occasions his staff cultured every inpatient at the medical center and found 20 percent of them *colonized* with VRE. That was incredible. The bug was proliferating wildly, not infecting most of its human hosts yet but residing in their gastrointestinal tracts for—well, who knew for how long? As long as the hosts maintained a strong immune system, they could live oblivious to the killer within them. Devastate the immune system, however, and VRE, resistant to all drugs, would be fully capable of causing a mortal infection. It was like a ticking time bomb inside one out of every five patients at the university medical center.

Shocked, Morris used his clout to demand that the university and affiliated VA hospitals at which he worked reduce their vancomycin use to all but the most urgent, life-threatening cases of active MRSA infection. Vancomycin was the ipso facto trigger for VRE, since without it, *E. faecium* would have no cause to become VRE. Without it, perhaps, resistant strains would even revert to susceptibility. The edict was met with angry protests from hospital doctors and nurses, who predicted many patients would die needlessly as a result of it, but Morris stood his ground. Intravenous vancomycin use dropped by nearly 59 percent. No greater incidence of deaths was reported.

That was the good news.

During this same period, Morris set up a system to test patients for VRE in the surgical intensive care and intermediate care units—the two areas in which the most VRE had bred—and isolate those patients found to be either colonized or infected with it. Nurses and doctors dealing with VRE cases were asked to abide by strict infection control measures. For starters, they had to wash their hands before and after each VRE patient contact. Throughout the contact, they also had to wear gowns and gloves. The measures were cumbersome, and expensive, and the bad news was that in the two years they were kept in place, the number of VRE infections in those wards declined hardly at all. Just as many new patients came in with VRE as before, likely as not colonized by the bug at other hospitals or nursing homes where lax infection control standards allowed it to spread by contact, and just as many patients in various other wards had it, too. Most depressing, compliance with the hand-washing rule, especially in the busy intensive care unit, averaged 30 percent—which was to say that when observed without their awareness, healthcare workers washed their hands properly with only one out of every three patients.

Afterward, when the hospital stopped screening all incoming patients to the ICU, rates of VRE shot up—and stayed up. Eight years later, VRE remained endemic at both of Morris's hospitals, like a guest who came for dinner and never went home. Out of 1,500 patients at the two hospitals, about 250 were colonized or infected with VRE whenever tests were carried out. Patients might check out cured of what had brought them into the hospital, but often they did so with VRE in their gastrointestinal tracts.

By the end of Morris's landmark study, VRE had spread through the mid-Atlantic states and was radiating out like a stain across the rest of the country. At the University of Virginia's medical center, Barry Farr, M.D., reported the incidence of VRE among enterococcal infections in his intensive care unit to be nearly 100 percent. In New York City, Barry Kreiswirth of New York City's Public Research Institute saw VRE emerge in about two dozen hospitals between 1989 and 1991; by 1999, Kreiswirth would report that 25 percent of all enterococcal infections in New York intensive care units were VRE. As the bug became endemic, many hospitals stopped trying to fight its spread by isolating infected patients and simply accepted that many of their sickest patients would succumb to it.

Dr. Don Low, chief of microbiology at Mount Sinai Hospital and Princess Margaret Hospital in Toronto, Canada, was one infection fighter who refused to be cowed by VRE. On February 19, 1996, when he got word that a patient at Mount Sinai had tested positive for VRE, he and his infection control director, Allison McGeer, took immediate and Draconian measures. The patient was whisked from an open ward to a private room. All medical equipment used on the patient — stethoscope, thermometer, and more — were kept in the room for use on him only. Everyone entering the room was required to wear not just gowns and gloves but double gloves: after touching the patient, a healthcare worker or family member would have to remove the outer layer before touching any surfaces. The measures seemed to come too late. In the days before the patient was found to have VRE, it turned out, he had infected at least one other patient.

Now Low and McGeer faced a torturous decision. Should they isolate the sixty patients on the two floors in question and subject the hospital to sizable costs and numerous hassles? Or just keep on as they had been, taking isolation measures for the "index patient," as medical jargon had it, and hope no one else had been infected? Not long before, they knew, a patient at the Toronto Hospital had transmitted VRE to thirty-seven other patients. In a Pittsburgh hospital, VRE's spread had already killed more than thirty patients in a liver transplant unit, and the hospital still had no idea how to expunge the bacteria. Grimly, Low and McGeer agreed to sequester both floors. That night, gowns and vinyl gloves had to be rounded up from medical supply stores for all sixty sequestered patients and their health-care workers, along with dedicated equipment and staff. All isolation rooms had to be disinfected to an extreme degree called "terminal cleaning" undertaken only for VRE. And all patients had to be given rectal swabs for VRE testing, with the results gleaned in hours by special methods rather than the usual three to five days. Another patient, then two, then five were all found to be infected with the bacteria. All, fortunately, were in reasonable enough health to be discharged and sent home, with stern instructions to home healthcare workers on how to contain their infections. The assumption was simply that if the bug did pass to healthy family members with strong immune systems, it would stand almost no chance of causing an infection.

By March 1, the remaining patients in isolation had all tested negative for VRE enough times that the outbreak could be declared over. For the patients, that was not a moment too soon: their prolonged confinement had left them highly irritable. But they were, at least, alive. Low and McGeer were relieved, though hardly sanguine. "This outbreak," McGeer said wearily at its end, "is just a sample of what's to come."

In one sense, the course of VRE strongly resembled that of methicillin-resistant *S. aureus,* or MRSA. Both bugs had appeared to originate in the intensive care units of large teaching hospitals, where many of the most chronically sick patients were treated. Both had spread to other wards in those hospitals, then to acute-care hospitals, community hospitals, and finally to long-term-care facilities and nursing homes. But MRSA had taken almost fifteen years to spread across the country—and remained susceptible to vancomycin. VRE had accomplished that spread in just five years—and was, of course, vancomycin resistant.

Why the ferocious speed of the spread? Surely, Morris felt, the fact that so much more vancomycin had been used in U.S. hospitals in the 1980s to combat MRSA was significant: the rate of this increase over the previous decades was literally 100-fold. That had to exacerbate the increase of VRE. *S. aureus* and *E. faecium* didn't often mingle, since the former usually hung out on the skin and in the nose and the latter colonized the gut, but *S. aureus* did occasionally visit the gut, too. Perhaps in the three decades since vancomycin had been used against *S. aureus,* enterococci routinely exposed to the drug had made their own efforts, unnoticed by scientists, to become resistant to it. Finally, some lucky enterococci had managed to import a resistance gene from some other strain or species of bug to protect themselves. Fred Angulo, the CDC's top man on food-borne pathogens, felt that resistance to a drug often took about that long to appear. Eventually, after trillions of genetic chances over tens of thousands of microbial generations, the bugs achieved a mutation that conferred resistance and then passed that resistance to their fellows. Once that happened, Angulo felt, an exponential increase in resistance occurred—the "spike" that had occurred first with MRSA and now with VRE.

In his university lab, Morris came up with another theory, one that linked VRE and *Clostridium difficile.* For centuries, another species of *Clostridium* had caused gangrene. The modern strain, which appeared only in hospital patients, caused what was known commonly as "antibiotic-associated diarrhea." It emerged because antibiotics, for all their benefits, inevitably knocked out much of the natural enteric bacteria, or flora, that a patient needed in the gastrointestinal tract. The change in antibiotic flora allowed *Clostridium difficile* to proliferate wildly and release toxins that caused diarrhea. In severe cases, the bacteria concentrated in the colon and caused the condition called pseudomembranous colitis, which caused bloody diarrhea, severe abdominal pain, and, in the worst cases, lethal shock. Morris hypothesized that the toxins released by *C. difficile* might make the gut more permeable, allowing VRE to infect the bloodstream. The drugs then used to kill *C. difficile*—the bug brought on by *other* drugs—might cause further damage to competing flora and foster the spread of VRE. Still, there had to be some other factor to explain its geographical origins and spread: Why the mid-Atlantic states first?

Maybe his hospital was just incredibly unlucky, Morris mused. Maybe it just happened to be the place where an enterococcal strain picked up the right combination of genes to become vancomycin-resistant. Almost certainly the bug had then spread, at least in part, because of lapses in infection control. But Morris thought there might be another contributing factor.

What if European visitors—both tourists and commercial travelers—had brought the first cases with them?[2] They tended to come to New York, Boston, and nearby Washington more often than to other American cities, because those cities were major business destinations and because they were closest to Europe. Perhaps, too, residents of those U.S. cities did more traveling in Europe than other Americans, also for reasons of business and propinquity. Either way, VRE genes might have originated in Europe. If so, this wouldn't have been attributable to the rampant use of vancomycin there—in fact, comparatively little was used. Rather, it would be due to the widespread, long-term use of a vancomycinlike antibiotic in livestock. Seeking answers, Morris turned to the ongoing, revolutionary work of a German geneticist named Wolfgang Witte.

Wolfgang Witte was not, perhaps, the likeliest sort to emerge as the Paul Revere of bacterial resistance in Europe. Tall and studious, with wire-rimmed glasses and slicked back dark hair, he was a young professor of bacterial genetics in 1983, raised and educated within the stern and dreary confines of East Germany. Yet because of the field he'd chosen, he wielded more clout than many of his Western counterparts. To the Communist government, understanding bacteria was

[2]The strains of VRE in Europe were mostly E. faecalis, while those in the United States were mostly E. faecium—different species within the genus enterococcus. But if resistance genes easily passed from one kind to the other, as most microbiologists believed, this hardly dented the theory.

a key to infection control, and infection control was a high priority: comrades might get little else from their government, but they did get good public health programs in order to stay fit enough to work. Witte got funding when he needed it, and his findings found a ready audience in the highest echelons of government. He also had a mordant sense of humor, which helped hardy souls endure life on the wrong side of the Berlin Wall. He had risen to his position at the small but prestigious Institute for Experimental Epidemiology, he explained, because his predecessor "went to pension, and the person who was to succeed him committed suicide. Thus I got my job."

As a respected researcher, Witte was allowed to read Western medical journals and correspond with foreign colleagues. So he was well aware of a growing controversy in the United States and western Europe about the use of antibiotics in animals not for therapeutic needs but as growth promoters.

As Orville Schell recounts in his classic book *Modern Meat,* the story of growth promoters had begun with a chance discovery in 1949 by one Thomas Jukes, director of nutrition and physiology research for the Lederle pharmaceutical company. Jukes was casting about for a natural source of the vitamin B-12 to supplement a standard livestock diet. By chance, he learned that the new antibiotic chlortetracycline, from the organism *Streptomyces aureofaciens,* yielded ample B-12 in wastes from the fermentation process used to produce it. Basically, the drug company put corn in a big fermentation vat, added the fungus, then let the concoction spoil and ferment until the antibiotic could be extracted from it. What remained was the fermented corn, which turned out to contain B-12. When the mash was used as feed, baby chicks did more than act perky. They grew faster, and fatter, than their fellow chicks.

Was this an accident? Jukes tried more of the mash on piglets and saw even more impressive results. Perhaps the antibiotic trace elements helped suppress harmful enteric bacteria, which in turn helped the animals grow. Perhaps they had some other effect. Jukes never managed to establish exactly why the mash worked as it did. All he knew was that it did work, no matter which antibiotic he used, and that it had a profound effect on agriculture in America. Livestock grew an average of 3 to 11 percent faster. And the livestock on growth promoters, Jukes found, also grew better—faster, fatter, and healthier—in tightly constricted spaces. This meant that far more livestock could be raised on a farm and that the farms, in turn, could grow much larger. Jukes, more than anyone else, could lay claim to the dubious honor of being the father of the vast farms known as CAFOs: confined animal feeding operations.

Astoundingly, no one in the agricultural or pharmaceutical industries ever demonstrated exactly *how* growth promoters promoted growth. Jukes felt it was just one of those miracles that science couldn't quite explain. Skeptics took a dimmer view. "If there was evidence they worked, the pharmaceutical industry would have provided that," observed the CDC's Fred Angulo. "They would have been falling all over themselves to show the evidence it worked." How then to account for the measurable increase in livestock growth from growth promoters? Possibly, growth promoters had merely offset bad hygiene; as hygiene improved, so did growth. So they worked, in a sense, but in Angulo's mind they were unnecessary.

Some consumer advocates went further. Growth promoters, they felt, were a marketing scheme—a brilliant means for drug companies to sell otherwise worthless, leftover mash. If so, this was a tragedy of epic irony: the greatest single invention in the history of medicine compromised by a huge exercise in American hucksterism, perpetrated by the very same companies that had developed human antibiotics! For by the late 1960s, evidence had begun to mount that these small, regular portions of antibiotics were a perfect prescription for precipitating resistance, first in the animals to which they were fed and then in the people who handled and ate those animals.

In 1969, a British government committee chaired by Professor M. M. Swann decided that anecdotal evidence was cause enough to recommend an across-the-board ban in the United Kingdom of any growth promoters that shared a class with human-use antibiotics. The committee had formed after an epidemic of multidrug-resistant *Salmonella* in people. Since livestock are almost the exclusive source of *Salmonella* in people, and the drugs to which this strain was resistant were from the same classes as the drugs used on that livestock, the connection seemed obvious enough to act upon. Soon after the ban was approved by the British government, levels of the strain, called DT109, subsided rapidly in animals and people.

In the United States, a landmark study on growth promoters was done in the mid-1970s by Dr. Stuart Levy. The dapper professor traded his signature bow tie and well-made suits for farm clothes to study the effects of a tetracycline antibiotic used as a growth promoter on 300 young chickens on a farm in Sherborn, Massachusetts. Within thirty-six hours of receiving oxytetracycline-laced feed for the first time, the chickens' intestinal *E. coli* went from tetracycline susceptible to tetracycline resistant. To Levy's surprise, the resistance broadened over the next three months to ampicillin, streptomycin, and sulfonamides—drugs that neither the chickens nor any other animals on the farm had ever been given. This was a deeply disturbing finding. Levy knew that sizable doses of a drug used therapeutically (to treat disease) sooner or later led the drug's bacterial target to become resistant to that one drug. But now it seemed that when the drug was fed to animals in tiny, subtherapeutic doses over time as part of their daily diet, it nudged bacteria just enough for them to gather resistance genes to all sorts of antibiotics. For the bugs, the whole exercise was like an aerobics class, making them sweat just enough, on a regular basis, to get in killer shape. And they were in shape to take on just about anything, whether they'd been exposed to it before or not.

The birds weren't the only victims: Levy found that the chicken farmers and their families acquired tetracycline resistance in their enteric *E. coli,* too. At the end of six months, that resistance, as in the chickens, broadened to other drugs. The farmers and their families had received none of those drugs. A control group of neighbors who had never visited the farm remained susceptible to tetracycline and the other antibiotics. The implication was clear: chronic use of antibiotics in animals led to multidrug resistance in people who ate or handled those animals.

Despite such dramatic evidence, the U.S. Food and Drug

Administration declined to ban tetracycline as a growth promoter. The politics were just too thick. The FDA, the very government agency that approves drugs for animals, received—indeed, still receives—its entire funding from the Agriculture Subcommittee of the Appropriations Committee of Congress. Virtually every member of that subcommittee represents a state with a strong agricultural industry. Ironically, Wolfgang Witte fared far better behind the Berlin Wall. When Witte appealed to the East German Council of Ministries to have tetracycline banned as a growth promoter, based largely on Levy's experiments, the council had no subcommittee to intimidate it and no private-sector industry to lobby it. The council decided to take tetracycline out, and that was that.

In East Germany, tetracycline was replaced by another growth promoter. This one, at least, was not in a class with any human-use antibiotic, so that the chances of it causing cross-resistance with human-use drugs seemed small to nil. That was a big improvement. It also gave Witte a neat chance to demonstrate animal-to-human transfer of resistance in as clear-cut a way as Levy had done with his chickens. At its introduction in 1983, resistance to the antibiotic nourseothricin was negligible. Two years later, it was found in *E. coli* in pigs. Not long after that, it spread to *E. coli* in the enteric flora of pig farmers, then radiated out to family members and neighbors. Was there, perhaps, some sliver of doubt as to where the local residents had picked up their resistance genes? If so, Witte erased it in 1989 by finding the same gene conferring resistance to nourseothricin in other enteric pathogens, including *Shigella,* which appear only in people. People weren't given nourseothricin, but the resistance gene for the drug had appeared in people, so it must have come from animals. Then it had appeared on other enteric flora that resided only in people. The circle of transfer was complete.

At about that time, Witte began to worry about the possible emergence of vancomycin-resistant enterococci. A clinical case of vancomycin-resistant *E. faecium*—the very kind that would soon appear with disturbing frequency in Glenn Morris's VA medical center in Baltimore—had shown up in East Berlin. So far, VRE had seemed to arise for one obvious reason: the overuse of vancomycin. Why, Witte brooded, would enterococci be under pressure to become VRE without the need to defend themselves against vancomycin, which was used quite sparingly in Europe? Didn't that shatter the most basic assumption about bacteria and drug resistance?

With the crumbling of the Berlin Wall, Witte's professional life brightened: his tiny institute merged with the Robert Koch Institute in Wernigerode, Germany, and he became a distinguished professor there. But to do research, Witte still put on his best knee-length rubber boots and visited local sewage plants. To his dismay, he found samples of the same VRE he'd seen in the clinical isolate in East Berlin. He hadn't thought the bug would be already ubiquitous in human feces. Witte could only assume that hospital use of vancomycin, while modest, had precipitated a rapid spread of VRE via the hands of healthcare workers and other environmental surfaces.

Strictly to form a control group for comparison, Witte then sampled the sewage plants of outlying towns with no hospitals. He expected to find no VRE at all, and so make a stronger case that the source of VRE was hospitals where vancomycin was used. Instead, he found more VRE—lots of it—in even the remotest treatment plants. Witte was mystified. Had lots of townspeople made recent trips to big-city hospitals? Witte soon confirmed they had not. Where, then, might the VRE have come from? That was when Witte remembered avoparcin.

Since its licensing in Germany in 1974, the growth promoter avoparcin had gone all but unnoticed by Witte and his colleagues. It was used only in animals, not humans. And unlike oxytetracycline, which shared a class with a widely used, broad-range antibiotic for human use, avoparcin seemed an insignificant analogue.[3] It belonged to a relatively small class called the glycopeptides. Chemically, it was related to a little-used drug in Europe called teicoplanin. But also to one other: vancomycin.

---

[3] An analogue is any chemically related compound—in this case a drug used in animals and sharing a class with a drug used in humans.

Like any good detective, Witte hurried over to the likeliest scene of the crime: the chicken and pig farms in Saxony-Anhalt and Thuringia where avoparcin was widely used. There he found animal fecal samples riddled with VRE. He tested the farmers' fecal samples, too, and those of their families and of people in the area who ate chickens and pigs from the farms. Ten percent of them were colonized with VRE. A particularly high incidence of VRE colonization showed up in people who ate a local specialty called *gehacktes*—raw minced pork spiced with onion, pepper, and salt. The raw pork was filthy with VRE, Witte found, as were pork sausages smoked on the outside but often rare, or almost rare, within.

Witte had connections to the medical school at Maggeburg University, and he talked them in to letting him perform an experiment. When the next truckload of frozen chickens arrived at the kitchen for the medical school's hospital, Witte was there to intercept it. He followed the shipment into the kitchen and watched the cooks thaw the chickens in ovens. The thawing process produced liquid, which Witte promptly tested: it was thick with VRE. The same strains found in the chicken liquid were then found in some of the hospital's patients. Good infection control measures kept the VRE from spreading, but to Witte the connection was clear: avoparcin provoked resistance in animal enterococci to the whole glycopeptide class, which meant the bugs were resistant to vancomycin, too.

Soon, Witte and a cadre of like-minded scientists found themselves in the fight of their lives, struggling to persuade the entire European Union to ban growth promoters. In that, they would be helped by the findings of a handsome French microbiologist who had just done the impossible: he had decoded the molecular mysteries of vancomycin-resistant enterococci.

4

THE GENETIC DETECTIVE

To reach Patrice Courvalin's office at the Pasteur Institute in Paris's fifteenth arrondissement, a visitor stops first at one of the wooden guardhouses that face each other across the narrow Rue Dr. Roux. Behind them rise the institute's massive, Beaux Arts limestone buildings, looking more like a museum than a medical research and teaching center. To anyone in the biological sciences, this is as awe-inspiring a sight as the White House, the Vatican, or the House of Lords is to others. A grateful public underwrote it after Pasteur's 1885 triumph in devising a vaccine for rabies. Contributions came from around the world—everyone from a postman in Normandy to the emperor of Brazil pitched in—to fund Pasteur's dream of an institute where doctors, chemists, and other professionals might trade their day jobs for lives in the lab as "Pastorians," pooling their research to help one another fight the world's worst diseases.

Within the grand façades of the original buildings, dreary stairways lead to dreary, high-ceilinged halls lined with doors that open in to cluttered laboratories. Courvalin occupies an airy corner office at the end of one such hall, with tall windows on two sides, a wall of medical texts and journals, and on another wall a large framed poster of a classic Jackson Pollock painting, *Blue Poles #11*. The arcing and swirling drips of the Pollock painting look like so many microbes under a microscope, though Courvalin says he just finds the image pleasing for what it is. Perhaps its energy is what attracts him. It's very much like Courvalin's own: restless and passionate, scattered and focused at the same time. Though he's now one of the institute's most distinguished members, Courvalin is still youthfully handsome and vigorous at fifty-six, with a full head of sandy hair and a sudden, beguiling smile, the kind of scientist who puts his feet up on his desk and still retains a bit of the rebellious spirit he had when he took an active role in the city-wide student demonstrations of 1968.

As cheerfully casual as he seems, Courvalin is, by anyone's count, among the world's top half-dozen leaders in bacterial research— some put him right at the top as the greatest microbe-hunter of his times for a string of dazzling discoveries about resistance that he and his lab have made over the last three decades. Of the many revelations emanating from Courvalin's lab, none is grander, or more elegant, than the genetic detective work that broke the code on how enterococci, against all odds, became resistant to vancomycin. From then on, no one would speak of *whether* bacteria would become resistant to a new drug, only *when*.

One day in 1986, a medical microbiologist named Roland Le Clercq wandered into Courvalin's lab with an agar plate and a baffled expression. Le Clercq had come to the lab for a while to do a Ph.D. on erythromycin resistance. As a routine matter, he always checked his bacterial strains against other antibiotics as well. Now one of his usual strains of enterococci appeared to be showing signs of resistance to vancomycin, potentially a very ominous development.

Courvalin was curious, but only mildly so. Over the last few years, other such reports of vancomycin-resistant enterococci had wafted into the lab. In every case, the reports had proved overblown. Either tests had been inaccurate or the bacteria had turned out not to be enterococci after all, but a similar-looking bug called *Leuconostoc,* which did have a naturally high degree of resistance to vancomycin, though no pathogenicity. "No, no," Le Clercq assured Courvalin, "these are enterococci." Courvalin shrugged. "Well, if you think so."

For the moment, Courvalin thought no more about Le Clercq's isolate: he had a lot to distract him. In the fifteen years since his student protest days, he had led a very different revolution: to look at bacteria on a cellular level. Instead of just streaking agar plates with various antibiotics and waiting to see if the bacteria grew resistant to them—the method Alexander Fleming had used to discover the antibiotic powers of a *Penicillium* fungus—Courvalin used new tools of molecular biology (the study of how genes function in cells) to see what bacteria were actually *doing* when they became resistant. He'd unzipped their double helixes of DNA to see where their genes of resistance lay. He'd cloned those genes into other bacteria to test their activity. Among other things, he'd learned that his colleagues were wrong in assuming genes only got swapped by bacteria within the same Gram class. Gram positives, it turned out, easily passed genes of resistance to Gram negatives—a revelation made even more alarming when he proved that those swaps occurred in direct reaction to human antibiotics. He was still working on transgram transfer, in fact, when he had his casual lab chat with Le Clercq.

If Le Clercq had pursued the implications of his strain, he would have been the one to publish the first paper on it: the discoverer of vancomycin-resistant enterococci. Instead, he went back to his Ph.D. work and put the telltale strain on a shelf. At first, out of propriety, Courvalin left it alone: it was Le Clercq's strain, after all. Then Eli Lilly sent the lab, as drug companies will, an experimental antibiotic to test against various bacterial strains. As a matter of due diligence, Courvalin asked Le Clercq if he could test his strain against the new antibiotic. *"Mais oui,"* Le Clercq said.

Courvalin dropped tiny paper disks impregnated with various antibiotics on a plate that contained Le Clercq's enterococcus: vancomycin, teicoplanin from the same glycopeptide class, and Lilly's new experimental drug. Then he went off to a medical conference. When he returned, the plate was gone. Courvalin asked the lab technician, an elderly woman who had worked at the institute for decades, if she'd seen his plate. "Oh, yes," she said. "I threw it out."

The technician explained that the plate was contaminated.

"Really?" Courvalin said. That seemed odd; Le Clercq was an extremely capable scientist.

"Yes," the technician went on, explaining that the population of Le Clercq's enterococci appeared not to be the only bacteria on the plate. There seemed to be a second population of some other bacteria. Whatever it was, it would throw off the results.

"Where's that plate?" Courvalin demanded.

"Over there, in the garbage can," the technician said.

With a muttered oath, Courvalin dug his hands into the trash. This was a far cry, he thought, from the Cartesian model of exacting research that every good French scientist learns. But when he fished out the plate, his irritation vanished. The technician's comment about two populations had given him an idea that the plate now seemed to confirm. Within a day, tests proved the discarded plate as important, in its way, as Fleming's left-out plate of *Penicillium*.

It was, Courvalin realized later, yet another example of Pasteur's dictum that chance favors the prepared mind. In an early experiment, he had shown that enterococci become resistant to tetracycline and erythromycin by means of plasmids. Courvalin had observed that the process involved a single population of bacteria splitting to become two. The first population, which was susceptible, was donating crucial plasmids to its daughter cells. The plasmids contained resistance genes, which were moved onto the daughter cells' chromosomes, so that they could be activated. The daughter cells formed a whole new population—one that was resistant to the drugs. It was as if a band of soldiers, nearly surrounded, had given their empty guns to their children because the children could sneak off to an ammunition depot, load the guns, and save themselves. For humans, that scenario would be the stuff of rare heroism. For bacteria, it was life, and death, as usual. Ultimately selfless, the bugs simply did whatever they could to perpetuate the species.

Here, just as then, Courvalin saw the plasmid-passing process at work. The only difference was that this time the enterococci were doing it against vancomycin. On Le Clercq's plate, just as he had suspected, there were indeed two populations, but both were enterococci. One was the old, susceptible population. The other was the new, plasmid-fitted, resistant one. Courvalin identified the plasmid that he thought conferred resistance, then proved his finding by transferring the plasmid's genes from a colony of resistant enterococci into one that was susceptible; the susceptible one became resistant.

With an article in the *New England Journal of Medicine* in 1988, Courvalin won the honor that Le Clercq had shrugged off, becoming the first scientist to demonstrate plasmid-mediated vancomycin resistance in enterococci—and to sound an alarm for every clinical doctor who treated enterococcal infections.[1] Here was the warning that all had dreaded and almost none had thought would come true. Vancomycin, the superdrug that vanquished Gram-positive bacteria, was not invincible after all.

Upon Courvalin's publication, researchers in microbiology labs around the world began vying to identify the cluster—or operon, as the vernacular had it—of genes that must be working in concert to create resistance. Courvalin set the pace, however. His team, led by Michel Arthur, elucidated which genes were involved and determined that five different ones, as the logic of science elegantly had it, were both necessary and sufficient to confer vancomycin resistance.

From their experiments, Courvalin's team could see that three of the five genes in the resistant enterococci appeared to work together to change the cross-links of the cell wall that vancomycin attached itself to—just how, the researchers weren't sure. The implications were horrifying—suddenly there seemed no drug to which enterococcus, and probably more virulent bugs, too, could not become resistant—but as a scientist Courvalin had to admire the intricacy of the bacterium's maneuver. He saw that the first of these three genes, which

---

[1] In fact, earlier that year, A. H. Uttley, C. H. Collins, and J. Naidoo published the first sighting of VRE, in the *Lancet*. But Courvalin was the first to explain how enterococci actually became resistant to vancomycin by means of plasmids, the more significant finding.

he called *vanA*, was almost identical to a gene in the susceptible bug. So nearly identical, indeed, that it seemed to perform the same cell-wall-building function adequately, if not quite as well, as the real McCoy. Yet it only appeared in the resistant bug. Adjacent to *vanA* on the resistance plasmid was a second gene that Courvalin called *vanH*. This one seemed to help *vanA* make a cross-link that was effective but just different enough from the original one to keep vancomycin from attaching to it. But then why the third gene, adjacent to the first two? It resembled no other known genes, so Courvalin called it *vanX*. He had no idea what this one's protein did and how it functioned, if at all, with the proteins of the first two genes.

Courvalin did think he knew what the fourth and fifth genes, *vanR* and *vanS,* did. He had observed that the new cross-linking system only kicked into gear when vancomycin appeared; otherwise, it lay dormant. That, too, made sense. Why add a layer of complexity to the cell-wall-building process when it wasn't needed? The extra effort took energy, and in the perfect Darwinian world of bacteria, energy was best conserved. But then the resistant enterococcus had to know when to turn its new system on and when to turn it back off. Courvalin theorized that *vanS* was the sensor, alerting the bacterium that vancomycin was in the neighborhood. He decided that *vanR* was the regulator gene that turned the system off and on.

As a microbiologist and a geneticist, Courvalin had completed his job. He'd shown that vancomycin resistance in enterococci was plasmid mediated, and he'd found the five genes responsible for it. But he still wasn't quite sure which genes did what. He knew which proteins they encoded and which amino acids, in turn, comprised the proteins. But he didn't know their biochemical functions or mechanisms of action. And *vanX* remained a mystery. As exacting as his vantage point was, there was one deeper level of understanding that would illuminate all those mysteries, and one somewhat impatient enzymologist in Boston who was eager to help.

Chris Walsh had read Courvalin's recent papers with more than a passing interest. As a professor of enzymology at Harvard Medical

School, he had made a career of studying cell wall biosynthesis, including the very enzyme that had prompted Courvalin's study and which now had given rise to the most awesome and complex mechanism of resistance ever seen in bacteria. He was fairly itching to get his hands on Courvalin's *vanA,* for starters. He wanted to purify the protein it encoded and then see if it really was a relative of the cell-wall-producing enzyme in normal enterococci.[2] That, in turn, might solve the puzzle of how the bacteria actually maneuvered their genes to create resistance: which proteins were substituted for the ones that made cell wall components that vancomycin recognized, and how the new ones built slightly different cell wall cross-links. But he had left messages for Courvalin several times over the last six months, and to his immense frustration the Frenchman hadn't called back.

Finally, Courvalin did call. He was flying to Boston, he said, to meet with a colleague. Would Walsh have time to talk? Walsh found time.

Courvalin sat warily at a blond-wood school table in Walsh's office as Walsh launched into his pitch. The Frenchman was the one who had the goods: Why should he share his gene with Walsh? In a hoarse voice that often trailed off, the straw-haired, bespectacled professor laid down his cards. First at the Massachusetts Institute of Technology, then at Harvard Medical School, he had built a reputation as one of the top enzymologists in the world. He had devoted his life to the chemistry of molecules in organisms—how they worked, why they did what they did. As the son of a scientist for the pharmaceutical giant Hoffman La Roche, he had always been interested in how his findings about enzymes might lead to new drugs. With the American's help, a drug might be developed for the looming threat of widespread VRE—a second-generation vancomycin, perhaps, or an entirely new drug. But a new drug would likely target one of the proteins that Courvalin's genes encoded, and there could be no hope of that until the chemicals were clearly identified. Walsh added that he could iden-

---

[2]Often, as in this case, proteins are enzymes. But the two are not always synonymous. Proteins are the bigger, more various group of which enzymes are one subset. So all enzymes are proteins, but not all proteins are enzymes.

tify the gene himself, eventually. By holding out, Courvalin would only slow the process. And what was the point of that?

Convinced, Courvalin sent Walsh the first gene, *vanA,* the one that produced the protein that had to be crucial because it appeared only in resistant enterococci. Walsh quickly confirmed that it was indeed a close relative, or homologue, of the original—similar, but not quite identical. Put back into enterococcus, it helped cross-link the cell wall, just as the original did, though not as effectively. Still, the mystery remained. If it did the same job, how was it different? Why did it do the job less well? And how did it confer resistance?

A young researcher in Walsh's lab named Tim Bugg began substituting different amino acids into the homologue. Perhaps the whole point of the homologue was that it could accept one or two *other* acids as substitutes for those that made up its amino acid pair. That very subtle difference might be enough to create a cell wall cross-link that vancomycin failed to recognize. In fact, Bugg was able to use any number of other amino acids as substitutes and show that they helped kick into gear the new, slightly different cell-wall-building process. Still, neither Bugg nor Walsh was sure which particular acid the bacterium used to beat vancomycin.

Now Walsh began studying *vanH,* the gene adjacent to *vanA.* It turned out to encode an enzyme that made an acid very similar, but not identical, to the original. And this acid did get taken up by *vanA.* But it was so bizarre! Walsh and Bugg had tested cross-links for years. None had ever taken this other kind of acid, so there was no reason to believe that that would occur when these two genes were put together. Nevertheless, it did. The gene product *vanH,* when expressed with *vanA,* created a new acid for *vanA* that could replace the original. To get this new combination, both *vanH* and *vanA* were needed. It was what geneticists called a "gain of function," an amazing phenomenon.

That left three more genes. Of those, Walsh unmasked two. The protein encoded by gene *vanS* was indeed the "sensor" that sensed when vancomycin was approaching and when it did acted as a motion-sensitive backyard light. And *vanR* was indeed the "regulator" that switched on the first genes—the genes that coded for enzymes that helped build new, vancomycin-resistant cross-links. *VanR*'s protein regulated this *only* when induced to do so by *vanS*. As soon as vancomycin entered its field of sensitivity, *vanS*'s protein signaled *vanR*'s protein to start up the whole extraordinarily complicated Rube Goldberg machine.

But what of *vanX*, the third of the three genes that seemed involved in cross-link building? Fitting this last piece into the puzzle became the work of the Third Man, a spry Cambridge University professor with wisps of white hair and a piercing look who did sort of resemble a spy. In an office so crammed with books and papers there was nowhere for a visitor to sit, Peter Reynolds began pondering *vanX* at Courvalin's invitation, and he realized intuitively what it must do. Proving his theory, however, took some time.

Reynolds was one of those Cambridge men who had arrived as an undergraduate and never left—his blurred image as a skinny boy in a class picture from 1960 hung on a wall of others in the biochemistry building, where he came to work every day as a senior lecturer. He had studied vancomycin when it had first appeared in the late 1950s—"your Ph.D. thesis is in that bottle," his professor had told him—and had been the first to show that it blocked cell wall synthesis. As a pure academic, he had gone on to other subjects and had spent much of the 1980s studying MRSA—he had been the first, as well, to elucidate the *mecA* gene that *S. aureus* employed to make itself resistant to methicillin—but vancomycin had remained a passion. As late as 1989, he had confidently written in a scholarly journal that enterococcus would never develop resistance to it.

Courvalin sent Reynolds a gene of *vanX* with a few pretty major clues. Michel Arthur, in Courvalin's lab, had noted that in a VRE cell, some twenty or thirty copies of the whole operon of resistant genes— all five members of the group that conferred resistance—were made. When each of the twenty or thirty *vanX* genes encoded a protein, the old cross-links in preexisting cell wall were changed to new ones. As a result, vancomycin had nothing to bind to, and the enterococcus was completely resistant to it. When Arthur removed some of the *vanX* genes from the equation, an odd thing happened: some of the old pathway remained intact. The fewer *vanX* genes there were in the enterococcal cell, the more of the old pathway that remained. The more of the old pathway that remained, the more susceptible the cell was. So it seemed the cell needed to get rid of all its old pathway even as it built its new one—and it seemed *vanX* was the tool it used to do that. But for a biochemist, that wasn't good enough. Reynolds had to know how *vanX* removed the pathway. He worked with Arthur to clone into vancomycin-susceptible enterococcus some operons that had *vanX* and some that did not. The ones that had it became 1,000-fold more resistant to vancomycin than those that lacked it, and the old cross-link in their cell wall was changed, indeed, to the new one. The difference was *vanX*.

Everything Reynolds had learned in his distinguished career suggested that an operon of resistance genes like this took millions of years to evolve. In that time frame, of course, it would have originated as a defense against soil bacteria. Yet if that were the case, the operon and its enterococcus would have come to have nucleic acids of the same composition. The acids of this operon and its host cell were chemically different. Reynolds was astounded. It seemed clear that the operon was a new arrival in enterococci with an obvious cause: the emergence and ever widening use of vancomycin as a human antibiotic.

As an expert in both VRE and methicillin-resistant *S. aureus*, Reynolds could see no reason why VRE should not now pass on its operon to the far more pathogenic, often lethal MRSA. They didn't even need to accomplish the transgram transfer that Courvalin had delineated: both were Gram-positive bugs. And while Reynolds himself had often stated that the two bugs wouldn't meet—MRSA inhabiting the skin and nose, VRE the gastrointestinal tract—none other than Courvalin had disproved *that* theory. They did meet, not often, perhaps, but enough, through surgery or other unusual circumstances, that the operon might easily be swapped.

Still, Reynolds was as startled as anyone else at the news, in late 1992, that a London researcher had decided not to wait for VRE and *S. aureus* to accomplish this swap themselves. He had done it himself.

In so doing he had created, in the eyes of his appalled colleagues in the field, the bacterial equivalent of Frankenstein.

Courvalin, as he tinkered with his genes of resistance, had wondered, as anyone would, if the operon might indeed be transferable from enterococcus to *S. aureus,* and in the most logical way he had tried to induce the swap. On an agar plate, he had put one colony of VRE and another of tetracycline-resistant *S. aureus,* then given the bugs time to swap resistance genes. Would the VRE pass the operon over to its beleaguered fellow bugs? Would the *S. aureus* pass its tetracycline resistance over to enterococcus? To see, Courvalin now added vancomycin and tetracycline. Time after time, the experiment failed. Though the colonies were close enough for cells of enterococcus to conjugate with cells of *S. aureus,* and for plasmids containing the resistance genes to be passed either way, they failed to do it. The *S. aureus* always died, vanquished by vancomycin.

In the research wing of St. John's Hospital for Diseases of the Skin in London, a white-haired, twinkly-eyed microbiologist named William Noble read of Courvalin's findings and chuckled. "I think he's cocked it up," Noble said to his fellow researchers. As much to tweak Courvalin as to advance the cause of science—out of "naughtiness," as he later admitted—Noble decided to coax VRE into transferring resistance to *S. aureus* his own way.

A generation before, Noble had written his own page in the history of *S. aureus,* coming at it as an expert in the ecology of the skin. In the early 1960s, he had waded into a fierce dispute between Australia and England over the origin of an especially virulent *S. aureus* strain known as 80/81. The strain had started in Australia and then spread around the world as a full-blown epidemic. Just at that time in Britain, the fabric industry had introduced cotton blankets; until then, all blankets had been made of wool. Because the blankets were associated with many of the cases, Australia declared that the *S. aureus* had spread on the cotton of these covers. Britain replied that Australia's wool blankets were to blame. Noble and a colleague named Roland Davies showed that both countries were right—and wrong. The *S. aureus* was being spread by exfoliating human skin. Whatever kind of blanket the skin particles fell into became a carrier of the infection as the particles touched new skin and the bacteria scrambled aboard. Until then, incredibly, no one had known that *S. aureus* lived on human skin.

From his thirty-odd years of studying this bug, Noble thought he knew where Courvalin had gone wrong. When the only players on the plate were the bugs and drugs that Courvalin had used, there simply wouldn't be time for *S. aureus* to make use of resistance plasmids from VRE before getting zapped by the drug. But Noble thought he knew of a back channel by which *S. aureus* might acquire vancomycin resistance from VRE after all. And he saw how he might demonstrate that channel, both in vitro and in vivo.

First, he found a rare colony of VRE that was susceptible to the antibiotic rifampin. (Most VRE were resistant to everything up to and including vancomycin, but there were exceptions.) Then he located a strain of *S. aureus* that was *resistant* to rifampin—but sensitive to almost everything else. It was not, as colleagues later muttered, MRSA: it was *sensitive* to methicillin, and to many other antibiotics, even penicillin. To Noble, that was a crucial element. If he did manage to transfer vancomycin resistance to *S. aureus,* he wanted to be sure he could kill it with all those other antibiotics. He intended to avoid exactly what he would later be accused of fostering: an *S. aureus* resistant to vancomycin and all other drugs.

Onto the agar plate, Noble put his rifampin-sensitive VRE, along with his rifampin-resistant *S. aureus.* Then he gave the bugs time to socialize without hitting them with any drug. Sure enough, the *S. aureus* passed its gene for rifampin resistance to the vulnerable VRE. At the same time, the VRE passed resistance genes of its own to the *S. aureus,* including, as it happened, resistance to erythromycin. Then Noble added rifampin and erythromycin. What he didn't add at this point was vancomycin because he knew vancomycin would kill his *S. aureus* before a neat little trick could occur.

Twenty-four hours later, both colonies were doing fine. Both, that is, had survived the onslaught of erythromycin and rifampin because both, thanks to their gene swaps, were resistant to both drugs. Now came the clever part. Noble had observed that the chunk of DNA that conferred erythromycin resistance also tended to have genes of resistance to vancomycin on it—like an igneous rock containing several different elements. With just a bit of grace time, he theorized, the *S. aureus* would be able not only to acquire those extra genes but to reconfigure its cell wall to be truly vancomycin-resistant. The proof came when Noble exposed his erythromycin-and-rifampin-resistant *S. aureus* to vancomycin. It was, indeed, wholly resistant to the drug. This was the back channel that Noble had theorized, the channel that Courvalin had overlooked by using vancomycin first.

Noble's article, published in a scholarly journal in 1992, caused a huge uproar of indignation, envy, and alarm. As news of the experiment spread, it quickly became distorted. What if Noble or one of his researchers had gotten the *S. aureus* on his hands, or in his nose? What if it had somehow escaped the laboratory? Then, they pointed out, it would surely be able to pass its vancomycin resistance to a strain of MRSA—by then, it would be all in the family. Noble waved away these fears: he had spent his whole career in a lab, and he knew how to take the proper precautions.

Doubtless he did. But his experiment remained hovering over the field as a horrifying prospect. It could happen—that was the lesson. Almost certainly, the impossible could occur: VRE could pass on vancomycin resistance to MRSA.

Four years later, from Japan, came the headline that all had dreaded: vancomycin-resistant *S. aureus* appeared to have arrived.

5

NIGHTMARE COME TRUE

Keiichi Hiramatsu scanned the chart with a strange mix of emotions. It detailed the case of a sixty-four-year-old man under treatment at the hospital affiliated with Hiramatsu's university lab in Tokyo. During or soon after an operation for lung cancer on December 3, 1995, the man had incurred a methicillin-resistant *S. aureus* infection in his lungs, itself a kind of pneumonia. At first, vancomycin had appeared effective against it. After the eighth day, however, the MRSA had begun to grow back. It was a sobering situation: the man seemed likely to die. At the same time, Hiramatsu felt a pang of excitement and awe—the pang a soldier might get on seeing an enemy army appear over the hill, equipped with exactly the new weapons he had feared they would have. It was a case he'd viewed as inevitable for three years now, ever since Noble's successful transfer of vancomycin resistance from enterococcus to *S. aureus*. Noble had shown it *could* happen, and the rampant proliferation of MRSA strains in hospitals throughout the world over the previous decade had made the threat an odds-on probability. Fully 50 percent of all *Staph* strains in U.S. hospitals were MRSA now, treatable only by vancomycin. Blitzed by that one drug again and again, some strain of the bug was almost certain to have a random mutation that gave it protection, then pass that mutation around or import some foreign piece of DNA from another species. Hiramatsu expected those scenarios; he didn't expect that the bug would devise an entirely new mechanism of resistance.

From his longtime work with MRSA, Hiramatsu suspected the patient's MRSA, unresponsive though it was to vancomycin, might still be susceptible to a beta-lactam drug such as penicillin or ampicillin in conjunction with a drug designed to knock out the bug's penicillinase enzymes. Pfizer, the U.S. pharmaceutical company, had sent him a promising new beta-lactam that combined ampicillin and sulbactam. He used that with arbekacin, an antibiotic that was available in Japan, but not the United States. In two days, the patient's fever subsided; in twelve days, the dwindling shadow of infection in his lungs was gone. Hiramatsu had saved the man's life. Now he studied a high-resolution photograph of the patient's sputum taken with an optical microscope before the combination therapy had begun. There it was: the bug he had known he would see one day. On a dish, it had a hue different from those of other *S. aureus:* more grayish than golden. And instead of being arrayed in identically sized clusters, its cocci appeared to be disparate in size, some larger than others.

Fascinated, Hiramatsu tested his sample to see if it had the genes that Courvalin and Walsh and Reynolds had identified as the mechanism of VRE.[1] But no: no *vanA, vanH,* or *vanX,* let alone *vanR* or *vanS.* What new trick of resistance, then, was this *S. aureus* using? The scientist's only clue was that, under a far more powerful electron microscope, the cells' walls seemed thicker than those of susceptible *S. aureus.* But what that meant, he couldn't yet say.

At the same time, Hiramatsu set about determining the bug's MIC, the minimum inhibitory concentration of drug needed to inhibit bacterial growth. The MIC was like the bar in a pole vault: the higher it was, the less susceptible, or more resistant, the bug was to the drug.

---

[1] He used a test called the polymerase chain reaction, which amplifies very small bits of DNA, making those bits easier to analyze and clone. The bits of DNA from the patient's sputum were compared to the enterococcal *van* genes (*vanA, vanH,* and *vanX*).

To his surprise, the isolate's MIC for vancomycin was 2 mg/L. That meant it took only two milligrams of vancomycin to inhibit all the bugs in the isolate he was testing. So the vancomycin should have worked: any bug with an MIC of 4 mg/L or less was defined as susceptible to the drug being tested on it. But the vancomycin had failed. Why? The obvious culprit was the patient himself: his immune system must have been too weak to help vancomycin do its job. But Hiramatsu had another theory.

For several years, Hiramatsu had been one of the few researchers in the world to subject bugs to a test more rigorous than the MIC, one called population analysis. The MIC considered the effects of a drug on 10,000 bacterial cells. That sounded like a lot. It wasn't. Hiramatsu looked at a drug's effect on exponentially more bacteria: with vancomycin, he used 10,000,000 cells. It was incredibly painstaking and tedious work involving many, many plates with many different concentrations of the drug, which was why so few other researchers, and no hospital labs, bothered to do it. But to Hiramatsu, the work was worth the time. Often, when he did it, he found small colonies of resistant bugs—like outlaws in hiding—within the larger groups of susceptible ones. Strictly by mathematical odds, you weren't likely to see those outlaws in a random pool of 10,000 cells. When you looked at a thousand of those pools, though, the odds began to rise.

To Hiramatsu, this situation justified a concept he called heterogenous resistance. An isolate with an MIC of 2 mg/L might seem susceptible at first look. But within it, tiny bands of outlaws in hiding made the isolate partly resistant even if the greater part was still susceptible. And in Hiramatsu's experience, a heteroresistant bug always seemed to become, sooner or later, homogenously resistant. Which was to say, *fully* resistant.

With his larger sample sizes, Hiramatsu found that when Mu3, as he christened the sixty-four-year-old man's strain of *S. aureus,* was subjected to higher and higher doses of vancomycin, its cells actually grew more and more resistant. The higher the dose of drug, that is, the more selective pressure was exerted, in this case enabling those tiny outlaw bands of resistant *S. aureus* to proliferate as the drug wiped out their susceptible brethren, until the MIC rose to 8 mg/L. That, as it happened, was the measure at which any bug was said to be intermediately resistant to the drug being used against it. In most clinical cases, a doctor could overwhelm intermediate resistance simply by blitzing the bug with more drug. And, to be sure, intermediate resistance was less of a threat than full-blown resistance, defined by an MIC of 32 mg/L or greater. But vancomycin was too toxic a drug to be given at ever higher levels; eventually it would do more harm than good. And if Mu3 was any indication, *S. aureus* had just learned how to match each higher dose of vancomycin with a higher level of resistance. This struck Hiramatsu as both appalling and important. And yet, when he submitted a paper on Mu3 as a first sighting of vancomycin-intermediate-resistant *S. aureus* to a Japanese medical journal, it was rejected. To the journal's editors, heterogenous resistance was a flaky idea. And Hiramatsu, despite some brilliant work to date, seemed to them to be jumping the gun for his own personal glory.

In his lab, Hiramatsu read the editors' letter with stung pride and keen frustration. They just didn't understand. The threat had already announced itself; all it needed was time to proliferate and cause widespread infection.

From what he knew of MRSA, Hiramatsu suspected that isolates with MICs of 8 mg/L or higher for vancomycin were already out there. They just hadn't been noticed yet, even if they'd caused treatment failures, because they hadn't been tested by population analysis. Grimly, Hiramatsu sent word in September 1996 to the microbiology lab of Juntendo Hospital, the one affiliated with his university lab.

He wanted to test every MRSA isolate they had.

At the international conferences he attended, Keiichi Hiramatsu struck all who met him as a courtly and elegant man, tall and dashing, with a full head of lush dark hair, given to well-tailored dark suits. But he had not become Japan's most prominent voice on drug-resistant bacteria without a steely determination. And for that, he credited his father. As a medical student at the start of World War II, Hiramatsu's father had been forced to abandon his dreams of becoming a medical scientist in order to become an army surgeon stationed in Manchuria. In the war's aftermath, he had been one of thousands of Japanese soldiers taken prisoner by Russian troops. Somehow he had managed to escape deportment to Siberia and embarked on an epic journey home by foot, selling nails and other detritus he scavenged from the streets to get money for food. The greatest goal, he would tell Keiichi as the boy grew up, was knowledge. Money or property could be taken away, but not knowledge. And what kind could be better, his father would add, than medical knowledge? Even the emperor had to put his life into a doctor's care. Keiichi became an internist, then switched to laboratory science to live out his father's unrealized dream.

When he issued his call for all MRSA isolates from the hospital's microbiology lab, Hiramatsu expected to find that one or another, when tested by population analysis, had an actual MIC of 8 mg/L or higher. But even he was taken aback by the hospital lab chief's reply. "Oh," she said, "we have about fifty isolates that might *all* be resistant to vancomycin." The strains, dating back to January 1996, had MICs of 2 to 4 mg/L defining them as susceptible, yet all had subpopulations with higher MICs than that. The lab chief had no case information to know if any of these strains had caused therapeutic failures, but she had a hunch that their resistance levels were subtly increasing; that was why she had kept them. To Hiramatsu, their heteroresistance was *very* significant. And the fiftieth one he tested, that of a four-month-old baby boy, had an MIC, by standard testing, of 8 mg/L—the accepted threshold of intermediate resistance.

The boy, Hiramatsu learned, had developed MRSA at the site of a surgical incision. In exposed, oozing pus, the infection might be spread easily on the unwashed hands of a doctor or nurse. Indeed, the boy had almost certainly contracted it from a patient in a neighboring bed who had been treated with vancomycin. The germs had either passed by hand from one bed to the next or hitched a ride on something that had come into contact with both patients: a pillow, perhaps, or a blanket.

Twenty-nine days of vancomycin therapy had failed to eradicate the boy's infection, Hiramatsu learned. On the contrary, it had made his MRSA more stubborn, pushing it up to the level of intermediate vancomycin resistance. Finally, his doctors had added arbekacin, the same drug that Hiramatsu had recommended in combination with a new beta-lactam for the sixty-four-year-old man. The boy seemed to respond to this vancomycin and arbekacin combo, enough to be taken home. Two weeks later, though, he was back in the hospital. The MRSA, now actually a vancomycin-intermediate-resistant *S. aureus*—soon to spawn the confusing acronym VISA—had reappeared in abscesses under the skin. The boy was in severe pain. His abscesses were debrided (removed by surgery), which caused more pain. After twenty-three days of infusion with Hiramatsu's combination-drug therapy, he recovered, went home, and relapsed again. Finally, after a third bout of antibiotics and more suffering for the boy, two years of sickness in all, the infection was killed for good. But Hiramatsu had no doubt the bacteria would return, in another human host, all too soon.

By then, Hiramatsu had confirmed by population analysis that the bacteria in the boy's abscesses had an MIC of 8 mg/L to vancomycin. He'd also compared Mu50, as he called the boy's strain, to Mu3, the strain found in the sixty-four-year-old man. The two strains were identical except in two respects. Mu50 had more resistant subpopulations than Mu3, and its cells' walls were nearly twice as thick. It seemed clear that the thickened walls of Mu50 must be part of a new resistance mechanism. Indeed, it appeared that Mu50 had evolved *from* Mu3, perhaps as a reaction to the ongoing use of vancomycin in those cases. But what exactly did the thicker walls do?

The answer would take time. For now, Hiramatsu sent his isolate to the CDC in Atlanta, Georgia. He wanted the CDC's confirmation of his MIC of 8 mg/L. With it, he submitted a report on Mu50 to the well-known British journal the *Lancet*. Surely, he thought, he had made his case at last. But an MRSA with an MIC of 8 mg/L, the editors said in so many words, seemed insignificant to them. The article was respectfully declined.

Hiramatsu's report appeared, belatedly, on July 11, 1997, in the CDC's weekly bulletin of outbreaks and new medical threats. Just three weeks later, a call came in to the CDC: the first-ever vancomycin-intermediate-resistant *S. aureus* infection, or VISA, in the United States had been discovered.

In a chain of command as elaborate as that of the U.S. Army's, Fred C. Tenover was, in effect, the CDC's five-star general in charge of the hospital pathogens laboratory branch, a category that included, sooner or later, pretty much everything. For all that, the office he occupied in one of the CDC's many nondescript buildings in downtown Atlanta was a windowless room hardly larger than a janitor's closet. You didn't join the CDC for big offices or high salaries. You certainly didn't come for fame or power, though you might nurture, as Tenover had done, a quiet ambition to wield influence of a sort: to help combat diseases that threatened America and the rest of the world.

Tenover was an unlikely warrior: balding, with a mild-mannered demeanor, he looked more like a lawyer or an accountant. His appearance, however, was as deceptive as his office. Tenover had come to science by way of Teilhard de Chardin, the French paleontologist and Jesuit priest who saw a confluence in science and religion, arguing that humans, having evolved from scattered cells, were evolving now toward a spiritual unity with God. Tenover's own marriage of theology and biology had led him to an encyclopedic knowledge of diseases on a molecular level and earned him the respect of every doctor, microbiologist, and hospital administrator with whom he now dealt. Under that Clark Kentish exterior was a bold microbe hunter, willing to work personally in the lab with some of the most dangerous germs on the planet.

Tenover was the official whom Hiramatsu had called when he sought CDC confirmation of his MIC for Mu50. Hiramatsu admired the American's expertise and knew that Tenover was gutsy enough to back a first finding of intermediate vancomycin resistance. In a vault-like lab with a sign on the door that read CAUTION—BIOLOGICAL HAZARD: DO NOT ENTER WITHOUT AUTHORIZATION just down the hall from his office, Tenover sat amid stacks of petri dishes, peered at the results of the MIC tests, and caught his breath. This was the result he had anticipated yet dreaded—a vancomycin MIC of 8 mg/L, just as Hiramatsu had said. The *S. aureus* had moved beyond the level where successful treatment with vancomycin could be assured. On petri plates, the colonies looked like a mixed culture of different cocci, except they weren't. As Hiramatsu realized, they were pure *S. aureus* clusters of different sizes. The real shock came when Tenover put the Mu50 culture under a powerful electron microscope. Compared to conventional *S. aureus*, the walls of these cells were *huge.* They appeared to absorb vancomycin like sponges and therefore prevent the drug from interfering with the cell's wall-building process within. Still, he thought, *Let's wait till we get an MIC of 32 mg/L*—the threshold of full-blown resistance—*before we start preaching doom and gloom.*

The call came one day in late July from Barbara Robinson-Dunn, director of microbiology at Michigan's public health department, whom Tenover knew well. "Hey, Fred," she said, a little out of breath. "We've got something really weird here that we think you ought to take a look at." A microbiologist named James Sunstrom at the Oakwood Hospital in Dearborne had called to report a *S. aureus* with an MIC to vancomycin of somewhere between 8 and 16 mg/L. Robinson-Dunn had passed it on to one of her senior lab scientists. The scientist had come back with an MIC of 8 mg/L. Robinson-Dunn was stunned by what she saw. This was a highly lethal, easily communicable bug that might be resistant to every antibiotic on the market. A bug right there in Michigan.

The patient with the VISA infection was a fifty-nine-year-old man who had been treated with several courses of vancomycin over a five-month period. For Robinson-Dunn and Michigan state epidemiologist Ken Wilcox, the overriding concern was obvious: had the VISA infection already spread to other patients? What she and Wilcox needed was an officer from the CDC's Epidemic Intelligence Service, right away, to track the bug and, if it had indeed spread, to help the state contain it.

"Got anyone in mind?" Tenover asked Bill Jarvis, acting director of the CDC's Hospital Infections Program.

"Yeah, I think I do," Jarvis replied. "I think I've got just the person."

At a little before 6:00 P.M. on August 12, 1997, Theresa Smith was at her newly assigned desk in the Hospital Infections Program at the CDC in Atlanta, Georgia. She had brought another box of her favorite medical articles to sort and arrange after hours—she had thousands of them, gathered over the years—and it pleased her to think that in a few more days, she would have her whole collection organized for the start of her two-year stint as an epidemic intelligence officer. But there was only so much that her articles could do to make her feel better. The truth, though she'd admitted it to no one, was that she felt lonely, very blue—and scared.

For the seventy or so applicants fortunate enough to be accepted into its ranks each year, the EIS offered more than a thorough grounding in public health. It was the Green Beret division of medicine, recruiting the best and brightest to serve as shock troops in the front lines of the war on lethal microbes. Most, like Smith, were doctors in their early thirties with a few years of clinical practice under their belts. They took pay cuts and uprooted their lives to be assigned either to one of the CDC's many surveillance groups in Atlanta or to one of a number of state health departments across the country tied in with the program. In their two years of service, the EIS members learned to be epidemiologists—medical detectives, in effect—tracking outbreaks and other emergencies to their pathogenic origins. The service had begun in 1951 as an early-warning system against biological warfare amid fears fanned by the burgeoning Korean War, but it had evolved into an emergency team to respond to epidemics of all kinds. EIS investigators had helped eradicate smallpox, solved the mystery of legionnaires' disease, reported the first cases of a rare pneumonia that became the AIDS epidemic, and identified a fast-acting respiratory killer as hantavirus.

At thirty-five, Smith was too old to be homesick and too experienced, as a practicing internist with a fellowship in infectious diseases, to feel out of her depth at the EIS. But she was also a shy, sensitive woman, devastated by the recent death of her mother—her best friend in the world. The sudden appearance of Michele Pearson, a physician and epidemiologist who served as Jarvis's aide-de-camp, filled her with apprehension.

"Looks like you hit the jackpot," Pearson said with a laugh.

Two days later, Smith was on a plane to Detroit. She'd made arrangements for her cat, gotten her ticket from the CDC travel office—and that was about it. There was no one she had to apprise of her sudden departure. She had a week's worth of clothes in her bag and, in her pocketbook, for good luck, a black-and-white photograph of her mother. The two cardinal rules that Jarvis had drummed into his recruits kept circling in her mind. *Don't believe anything anyone tells you,* he had said. *And don't come home until you've solved it.*

Early the next morning, Smith met Ken Wilcox and Barbara Robinson-Dunn in her hotel lobby. The three drove together over to the Royal Oak office of Dr. Cosmo Cruz, a nephrologist who was the VISA patient's primary doctor. There, too, was Dr. Jeffrey Band, the head of infection control at William Beaumont Hospital, the institution where the patient had been cared for most recently.

After brief formalities, Cruz laid out the case. The patient had diabetes mellitus, hypertension, chronic kidney failure that had required dialysis for more than five years—and, as if that were not enough, metastatic lung cancer. He was just the kind of weakened human specimen that a few billion super-resistant *S. aureus* were liable to use as a living petri dish before attacking other members of the species. The previous February, the patient's peritoneal (lower abdominal) fluid had grown MRSA, probably due to infection from the catheter to which he was tethered for peritoneal dialysis, and so a fourteen-day course of vancomycin had been initiated. The infection had disappeared, then reappeared. Over the next months, the patient had received four more courses of vancomycin, mostly at home with the help of healthcare workers, as the infection kept coming back. He was, in short, a very sick man, who on July 19 suddenly got much sicker when his peritoneal fluid grew VISA.

Fortunately, the patient's VISA isolate had proved susceptible to a few second-line antibiotics not usually used with MRSA. The susceptibility profile jibed with Tenover's lab analysis, down in Atlanta, that this VISA was slightly different from the Japanese strain but that its apparent mechanism of resistance, the swollen cell wall, was the same. Smith agreed it was time to try the most logical back-up drug, rifampin, and stand ready to add trimethoprim-sulfamethoxazole. From a public health perspective, the more pressing issue was containment. Smith would need detailed lists, if possible, of virtually everyone who had come in contact with the patient. All would have to be tested for possible spread, or carriage. The hospital's methods of infection control would also have to be scrutinized.

The patient had been so sick for so long, Smith learned, that his contacts had been limited almost exclusively to his wife and his doctors and nurses. The concentric circles, to use Jarvis's term, were minimal. Even so, the list would grow to more than seventy-five names. The question of hospital procedure was more delicate. If she concluded that hospital administrators had not covered all the bases of infection control at William Beaumont, Smith would have to advise them, gently or not, which policies to change immediately, a potential cause of conflict that all EIS officers learned to dread.

Jeff Band was calmly adamant that every precaution had been taken. Smith felt she could believe him. "It's your hospital," she said quietly, "and it sounds as if you've done everything there is to do."

The next day, with lists provided to her, Smith began the tedious work of going from one doctor's office to the next, testing every contact named. She swabbed the hands of each one with a sterilized wipe, and with a Q-Tip-like surgical stick she swabbed each contact's nostrils. Those were the sites where *S. aureus* was likeliest to reside. The wipes and the swabs were put in plastic cups, marked with the contact's name, and added to a growing collection.

On Saturday, after soliciting Cruz's help as a go-between, Smith also paid a visit to the patient at his home. Debilitated and depressed after his many bouts of infection and therapy, he now felt angry and keenly frustrated. Until July 19, his home healthcare workers had cared for him without any protective clothing, unaware, as it turned out, that he had recurring MRSA infections. Now, every time they entered his room, they had gowns and gloves and surgical masks on. So, too, did the man's wife, under strict orders from Band and the patient's other infectious disease specialist, Dr. Marcus Zervos. The precautions made the patient feel even more like a pariah than he had felt before—and even more lonely. When, he wanted to know, would all these precautions end?

Gently, Smith explained why the precautions were necessary and why the patient's case had become so important. She thanked him, and his wife, for letting her conduct her investigation. When would he be rid of this infection? "Soon, I hope," she said. There was nothing more she could say, but her presence did seem to mollify the patient and his wife. She wasn't some government bureaucrat acting all officious, they noted. She was more like a shy farm girl, with her long dark hair, kaleidoscopic blue eyes, and soft pink cheeks.

Four days later, Smith had ticked off most of the contact names on her list, knocking on doors and conducting interviews like a politician in the heat of a campaign. A few of the contacts were employees on vacation; they would have to be done later. But of the seventy-five contacts she swabbed and tested, none had VISA. This was a vast relief, though not too surprising, given that the patient's infection, unlike the Japanese baby's suppurating wound infection, was an internal one. Few healthcare workers had touched the dialysis catheter from which the VISA bugs could have spread. *S. aureus* had been found in 25 percent of the hand cultures taken—thirteen out of fifty-one—and 15 percent of the nostril cultures, but that was about what Smith had expected to find in any such random sampling. This, after all, was susceptible *S. aureus,* an all-too-common inhabitant of the human skin and nostrils. But the implications were unsettling nonetheless. Healthcare workers were likely spreading this *S. aureus* from patient to patient. Any patients with open cuts were vulnerable, at the least, to a *S. aureus* infection that drugs *could* kill. But the more drugs that were used, over time, the more likely the bugs were to become resistant. And if a VISA strain *had* managed to find its way to one of those workers' hands or nostrils, what was to stop it from spreading as easily as those susceptible strains?

In her hotel room, Smith laid out all the charts and began compiling a report from them, like a journalist on deadline, that would appear in the CDC's next *Morbidity and Mortality Weekly Report*. Occasionally, she let her gaze shift to the black-and-white photograph of her mother she had slipped into the inside cover of her organizer. So much of why Smith was there, in that hotel room, had to do with her.

As one of six children born to a family of doctors and nurses, Theresa Smith had grown up in Des Moines, Iowa, with a strong sense of public duty. You didn't aspire to be rich, or famous, or powerful. You found the best way to apply your God-given talents to the service of others. For Theresa that commitment was dramatized by her mother's decision to care for her second-born son when the boy, at age two, was stricken with polio. The year was 1950, just before the Salk vaccine became available. Theresa's mother chose to care for her son Phil at home, knowing full well, as a nurse, how easily the polio virus could be passed through contact. Despite her best professional efforts, she contracted the virus herself. Theresa, fifteen years Phil's junior, grew up seeing her mother in a long leg brace, with crutches; her brother could walk unaided but resorted to knee-to-ankle braces for support. Theresa went through medical school at the University of Iowa, then stayed in Des Moines to do a fellowship in infectious diseases. Now, as an EIS officer, she was doing the work she'd aspired to for so long. On her desk in the hotel room, she propped up a five-by-seven black-and-white photograph of her mother, at age seventeen, standing against a tree: a beautiful and demure young woman with strong and healthy legs exposed below a knee-length woolen skirt.

On Wednesday, almost as an afterthought, Smith decided to call Michele Pearson in Bill Jarvis's office. She found it a bit odd that Pearson hadn't called her in the last several days.

"I wondered when you would check in," Pearson said, bemused. "Don't you know how to use the Async system?"

Smith cringed. Amid the pressures of the last several days, she had completely forgotten that a new voice mail system had been instituted by the CDC. Pearson had left her several messages; Smith had received none of them.

"Why, what's up?" Smith asked.

"Well, you're getting on a plane tomorrow," Pearson said. "But not back to Atlanta."

"What do you mean?"

"You're going to New Jersey," Pearson told her. "We've just had a second report of VISA."

When Smith arrived at her next hotel in Camden, New Jersey, at 11:00 P.M. on Thursday, August 21, 1997, there was a message waiting for her from Dr. Colin Campbell of the New Jersey state health department. Campbell wanted her to call him on arriving, no matter how late the hour.

Briefly, Campbell filled in the picture Pearson had provided. On August 6, a blood culture from a sixty-six-year-old male patient at Our Lady of Lourdes Hospital had been drawn and submitted for testing in the hospital's microbiology lab. It had produced an MIC of 8 mg/L for vancomycin, nicking the threshold, as had the Michigan case, of intermediate resistance. The tests had been done again. Then different tests had been done. There was no doubt about the result. The patient was a Catholic priest with a large family of siblings, nephews, and nieces, who had taken turns caring for him over the several months that his condition had steadily deteriorated. The hospital's top administrators and the man's family were anxious that the case remain secret. The last thing they wanted was national headlines about an incurable superbug at Our Lady of Lourdes Hospital.

Early on Friday, August 22, Smith met with the patient's doctors and tried, once again, to make sense of a baffling situation. News of a second case raised an obvious question: Had the bug somehow spread from Michigan to New Jersey? Almost certainly, the answer was no. Fred Tenover's lab at the CDC had just confirmed that the isolate was indeed VISA, but also that the New Jersey strain was slightly different from the Michigan one, though both were variations of Mu50. And while Smith's investigation might prove otherwise, there appeared to have been no human contacts common to the two cases, which were, after all, hundreds of miles apart. Smith wondered, however, if she might be seeing a field effect. A smoker's lungs subjected to daily doses of carcinogens might as likely develop two or more separate malignancies as one, because the whole field had been affected equally. Similarly, *S. aureus* bacteria in the bodies of geographically disparate patients, if subjected to identical onslaughts of vancomycin, might become resistant at the same rate, entirely independent of each other. It was both unlikely and possible. In fact, the priest had received vancomycin for eighteen weeks overall—exactly the same length of time as the patient in Michigan.

Of course, if this *was* the start of a field effect, it could mean that patients all over the country undergoing long-term vancomycin treatment were on the verge of developing VISA, too.

From Drs. Gary Burke and Rob Williams, Smith learned that the priest's health had begun to decline two years before, when congestive heart failure forced him to undergo bypass surgery. By February 1997 he was forced to reenter the hospital with diabetes, deteriorating kidney function, and an autoimmune reaction that decreased the platelet count in his blood. It was during that stay that he'd first become infected with MRSA and undergone his first course of vancomycin. Within days, he suffered complete kidney failure and had to go on dialysis. Very likely, the drug caused the renal failure. As likely, in turn, the renal failure may have precipitated the VISA: without healthy kidneys to flush it out, the drug lingered in the priest's system longer, giving his MRSA that much more time to become accustomed and, eventually, resistant to it.

When he came off dialysis, the priest had gone home to the gabled rectory house where he lived beside his parish church in Camden. But his MRSA kept coming back, requiring home healthcare and more courses of vancomycin. That spring, he had possible pneumonia, severe back and shoulder pain, and breathing difficulties, as well as the recurring MRSA. He was also losing weight at an alarming rate. On July 25, when he was admitted to the hospital, he weighed eighty-two pounds less than at his admittance in February. His kidney function was worse again, and his heart function was faltering. A blood culture taken on that day measured 4 mg/L resistance to vancomycin, worrisome enough to warrant follow-up cultures. On August 8, he left the hospital again, determined to improve with intravenous vancomycin at home. It was in his follow-up cultures that the MIC of 8 mg/L to vancomycin was revealed.

For the moment, the patient was still at home, under full-time medical care, on vancomycin and a second antibiotic, gentamicin, to which his VISA strain had proved susceptible. The combination therapy, prescribed by Rob Williams, actually seemed to be working: the MIC was going down. Smith asked if the home healthcare workers were wearing gowns and gloves and discarding or sterilizing every object that came into contact with the patient. They were. She asked what the hospital was doing to shield its other MRSA patients from possible infection. For starters, she said, were they isolated?

There was an uncomfortable pause. "We aren't isolating MRSA cases at this time," explained Lori Boschetto of the hospital's infection control team. They'd done it a few years back, she explained, and observed that isolating the patients made no difference. "The numbers didn't change, because there were so many coming in from the community. Also, we weren't having internal transmission problems."

In light of the VISA case, Smith suggested, perhaps isolating the MRSA cases was a sensible precaution. But she made clear that the decision should be made by the infection control team.

Already, Smith could see that the list of patient contacts to be tested would be longer than that for the Michigan patient, because of all the different medical offices the New Jersey priest had visited over the last year or more and all the relatives who had come to care for him. Eventually, it would grow to 156 doctors, nurses, other healthcare workers, family members, friends, and hospital patients who had had contact with the patient. Boschetto and Catherine Crain, also of the hospital's infectious disease staff, volunteered to help Smith track down and test contacts. At the same time, they brought up the other problem Smith would have to address: the hospital staff was in a state of panic.

Smith agreed to moderate an open staff meeting early the next week. More immediately, she called the seventeen members of the priest's extended family who lived in the Camden area and had taken turns caring for him. She asked one of his two sisters to relay to the priest her request to meet with him, and when an invitation came back, she asked the relatives to meet her at the parish house so she could take their cultures while she was there. She made a point of arriving at the gabled building before the relatives did. She found the priest debilitated, as she had expected, but still possessing a puckish charm and able to sit up in the living room. With a smile, he waved her into the kitchen, where she had asked to set up a laboratory for testing cultures. And he kept up a steady banter of wry remarks as the relatives arrived one or two at a time. But she could see, by the time the testing was done, that the visits had taken a toll. "I'm going to give you back your house now," she told him at last. As she leaned over to say goodbye, in her gown and gloves, she told him that she, too, was Catholic, and asked him for his blessing. It was the one thing, she thought, that he could still give. And she could see that it pleased him to give it.

A lot of anxious faces looked up as Smith arrived at the open staff meeting and took the floor. Her presence was disarming—the nurses had half-expected a team of CDC officers in white space suits—but the group had hard questions to ask, and they wanted answers. Had their hospital exposed them to a lethal infection that might already be festering in their bloodstreams? Were their families at risk?

Smith tried to convey a calmness she didn't entirely feel. She explained that VISA was just like ordinary *S. aureus,* only a lot more resistant. Everyone treating MRSA patients would have to wear gowns and gloves, more to protect the patients than themselves, and, if the priest with VISA was readmitted, special precautions would have to be taken. He would be put in the hospital's reverse isolation room, a unique chamber designed for tuberculosis patients: fresh air circulated into the room, then circulated out through a high-efficiency filter that could remove particles as small as bacteria. In addition, he would need a dedicated nursing staff. And everyone entering the room—doctors, nurses, family members, and friends—would be required to wear masks, plastic gowns, double gloves, and shoe covers, all of which would be discarded into a designated linen bag on departure. Even the patient's lab specimens would have to be sent down in a special way. With all that, she said, no one should be in any more danger than from MRSA. After all, *S. aureus* wasn't known to be an airborne organism. Though as another precaution, she asked all staff members to submit bacterial specimens from their hands and noses.

"How long are we supposed to keep this a secret?" one person demanded.

That was a hard one. Somehow, the news had leaked that a superbug had emerged somewhere in the Camden area. Newspaper stories were already fanning fears of a potential nightmare for hospitals, spread among already vulnerable patients. Williams wanted the hospital to hold a news conference right away, but the administrators had ruled against it. Apparently they were hoping the hubbub would just subside.

Two days later, to the alarm of many in Our Lady of Lourdes, the parish priest was readmitted. Once again, his kidneys were failing; he needed dialysis. In he went to his tomblike private room, escorted by the nurses who had volunteered to care exclusively for him.

With so many healthcare workers aware of the priest's condition, news of his whereabouts was bound to leak—and on September 3, it did. Television news vans gathered in a buzzing cluster outside the hospital, demanding answers from all who walked by. Furious and fearful, the hospital's administrators scheduled an emergency news conference. Smith was asked to preside. She spoke in a calm voice to a sea of reporters—she could never have imagined, sitting scared and alone at her desk at the CDC two weeks before, that she would be forced to do this, much less that she would be capable of it—and fielded questions that bordered on the hysterical. *How many patients are likely to die? Is the public at risk? Is this like the Ebola virus?*

Four days after the news conference, a bone-tired Smith flew back to Atlanta. She had overseen the culturing of all 156 contacts

Hiramatsu's report appeared, belatedly, on July 11, 1997, in the CDC's weekly bulletin of outbreaks and new medical threats. Just three weeks later, a call came in to the CDC: the first-ever vancomycin-intermediate-resistant *S. aureus* infection, or VISA, in the United States had been discovered.

In a chain of command as elaborate as that of the U.S. Army's, Fred C. Tenover was, in effect, the CDC's five-star general in charge of the hospital pathogens laboratory branch, a category that included, sooner or later, pretty much everything. For all that, the office he occupied in one of the CDC's many nondescript buildings in downtown Atlanta was a windowless room hardly larger than a janitor's closet. You didn't join the CDC for big offices or high salaries. You certainly didn't come for fame or power, though you might nurture, as Tenover had done, a quiet ambition to wield influence of a sort: to help combat diseases that threatened America and the rest of the world.

Tenover was an unlikely warrior: balding, with a mild-mannered demeanor, he looked more like a lawyer or an accountant. His appearance, however, was as deceptive as his office. Tenover had come to science by way of Teilhard de Chardin, the French paleontologist and Jesuit priest who saw a confluence in science and religion, arguing that humans, having evolved from scattered cells, were evolving now toward a spiritual unity with God. Tenover's own marriage of theology and biology had led him to an encyclopedic knowledge of diseases on a molecular level and earned him the respect of every doctor, microbiologist, and hospital administrator with whom he now dealt. Under that Clark Kentish exterior was a bold microbe hunter, willing to work personally in the lab with some of the most dangerous germs on the planet.

Tenover was the official whom Hiramatsu had called when he sought CDC confirmation of his MIC for Mu50. Hiramatsu admired the American's expertise and knew that Tenover was gutsy enough to back a first finding of intermediate vancomycin resistance. In a vault-like lab with a sign on the door that read CAUTION—BIOLOGICAL HAZARD: DO NOT ENTER WITHOUT AUTHORIZATION just down the hall from his office, Tenover sat amid stacks of petri dishes, peered at the results of the MIC tests, and caught his breath. This was the result he had anticipated yet dreaded—a vancomycin MIC of 8 mg/L, just as Hiramatsu had said. The *S. aureus* had moved beyond the level where successful treatment with vancomycin could be assured. On petri plates, the colonies looked like a mixed culture of different cocci, except they weren't. As Hiramatsu realized, they were pure *S. aureus* clusters of different sizes. The real shock came when Tenover put the Mu50 culture under a powerful electron microscope. Compared to conventional *S. aureus*, the walls of these cells were *huge*. They appeared to absorb vancomycin like sponges and therefore prevent the drug from interfering with the cell's wall-building process within. Still, he thought, *Let's wait till we get an MIC of 32 mg/L*—the threshold of full-blown resistance—*before we start preaching doom and gloom*.

The call came one day in late July from Barbara Robinson-Dunn, director of microbiology at Michigan's public health department, whom Tenover knew well. "Hey, Fred," she said, a little out of breath. "We've got something really weird here that we think you ought to take a look at." A microbiologist named James Sunstrom at the Oakwood Hospital in Dearborne had called to report a *S. aureus* with an MIC to vancomycin of somewhere between 8 and 16 mg/L. Robinson-Dunn had passed it on to one of her senior lab scientists. The scientist had come back with an MIC of 8 mg/L. Robinson-Dunn was stunned by what she saw. This was a highly lethal, easily communicable bug that might be resistant to every antibiotic on the market. A bug right there in Michigan.

The patient with the VISA infection was a fifty-nine-year-old man who had been treated with several courses of vancomycin over a five-month period. For Robinson-Dunn and Michigan state epidemiologist Ken Wilcox, the overriding concern was obvious: had the VISA infection already spread to other patients? What she and Wilcox needed was an officer from the CDC's Epidemic Intelligence Service, right away, to track the bug and, if it had indeed spread, to help the state contain it.

and sent them on to the lab, which had determined that not one of them had a VISA infection, though 34 percent of them tested positive for *S. aureus*. Was it just luck, she wondered wearily, that no one—especially the healthcare workers who had dealt with the priest on a regular basis—had caught VISA, given how easily *S. aureus* spread? Was it entirely due to the infection control measures instituted at the hospital? Or was the bug somehow unable to infect other human hosts? Perhaps the bugs' response to vancomycin, the thickened cell walls, kept them alive but diminished their ecological fitness. Perhaps, though the mechanism worked, it rendered them as sick and hapless as their hosts.

In his private room at Our Lady of Lourdes Hospital, the priest did his best to stay hopeful as September ticked by. Every day, his sisters and other relatives continued to visit, though their visits were hampered somewhat by the masks, gloves, and gowns they had to wear. In between visits, the priest gazed at the get well cards that covered the walls and listened to Broadway scores or Irish folk music on his CD player. At night he often rang for one of his favorite nurses, Karen Stronsky, and asked her to put on the song he loved best, "Lady of Knock, Queen of Ireland." (The title was one of the 325 recognized names for the Virgin Mary.) She would sit in the room listening to it with him, and when it was done, she would ask, "Do you want to hear it again?" She could see, as the month unfolded, that his hope was ebbing with his physical strength and that he had accepted he would never leave the double-isolation room alive. "You're ready to be with the Lord, aren't you?" she asked him at one point.

"Yes, I'm ready," came the feeble reply.

On October 4, the priest expired—his last hugs, from his sisters, through masks, gowns, and double gloves. His blood was still free of the VISA infection; the cause of his death appeared to be a fungus. But the fungus had taken hold because his natural bacterial flora had been decimated by the antibiotics given to rid him of the VISA. A circle of interlinking ills, in the end, was what had killed him, though, without the VISA to exacerbate all the others, in all likelihood he would have survived.

The VISA patient in Michigan died three months later, though he, too, was free of VISA when he died; in his case, the direct cause of death was lung cancer.

Two bad quirks out of the blue? So it seemed, for a matter of months. And then, just as the scientists at the CDC began to think it all a bad dream, the new cases began to come in.

6

TWO NOT-QUITE-MAGIC BULLETS

If the first two American victims of vancomycin-intermediate-resistant *S. aureus* had, by a strict medical definition, survived their infections, the third most certainly did not. On March 20, 1998, a seventy-nine-year-old retired New York City narcotics detective who lived in Westchester County woke up feeling feverish and disoriented. He was in ill health already: failing kidneys had forced him to undergo dialysis three times weekly for the previous eighteen months, complicating a history of heart disease and hypothyroidism. Then, in January, he had incurred a methicillin-resistant *S. aureus* infection from an infected catheter. That had put him on a course of vancomycin for several weeks before coming home. Now, at his daughter's insistence, he checked in to the United Hospital Medical Center in nearby Port Chester, where he was put on vancomycin again. At the same time, a culture of his blood was rushed over to Sharon Rotun, head of the hospital's microbiology lab. Twelve hours later, before Rotun even got the results, the retired detective was dead. Just what killed him remained unclear, in part because the detective's family forbade an autopsy to be done, in part because the patient had had so many complications. Clearly, though, VISA contributed to his death.

Rotun, as startled as the lab scientists in the Michigan and New Jersey cases, repeated her tests before confirming the first case of VISA in New York State. As in the Michigan and New Jersey cases that Theresa Smith had investigated for the CDC, the MIC for vancomycin in the Port Chester case was 8 mg/L. Fred Tenover, studying the isolate in his windowless lab at the CDC, felt relieved, at least, that the MIC was no higher than that—still a far cry from the 32 mg/L that signaled full-blown resistance. And yet the patient had died. Was it possible, Tenover wondered, that an MIC of 8 mg/L was all *S. aureus* needed to fend off vancomycin in other clinical cases, too?

In other words, had full resistance by *S. aureus* to vancomycin—clinical resistance—already arrived?

Ominously, too, DNA analysis confirmed that the strain of MRSA that had escalated to VISA in the Port Chester man matched eight MRSA isolates culled from different hospitals in the New York City metropolitan area. At New York's Rockefeller University, Alexander Tomasz and Krzysztof Sieradzki demonstrated that all eight strains could be nudged in vitro to develop intermediate resistance to vancomycin. An entirely reasonable conclusion was that VISA seemed likely to spread through those metropolitan hospitals and beyond.

A successor to vancomycin was clearly needed. So, when a new drug succeeded just two months later in knocking out a lethal *S. aureus* infection in a patient in which vancomycin had failed, Tenover was both grateful and amazed.

Sixteen-year-old Teresa Miltonberger's ordeal began on the morning of May 21, 1998, when freshman Kipland Kinkel opened fire in the crowded cafeteria of Thurston High School in Eugene, Oregon, just before the start of classes. Two students died and twenty-two were wounded. Teresa was shot twice in the head—one bullet passing from one side of her forehead to the other, a second entering the back of her skull—and once in the leg. A health instructor's mouth-to-mouth resuscitation saved her from dying at the scene, but doctors at nearby Sacred Heart Medical Center judged her chances of survival to be just 10 to 20 percent. During an initial surgery, two displaced skull fragments were successfully removed from Teresa's head, but an infected abscess formed in her brain, one soon diagnosed as MRSA. Teresa was put on vancomycin, only to incur a hypersensitive reaction to the drug called red man's syndrome: her skin literally turned red, and she spiked a fever, began throwing up, and was rushed into intensive care. Her parents were told that Teresa had to be taken off vancomycin right away. What, then, would she be put on instead for her life-threatening MRSA infection? "We've got nothing else," one of her doctors said gently.

Before the day was out, however, another of her attending physicians placed an urgent call to the U.S. headquarters of French drug company Rhone-Poulenc Rorer in Collegeville, Pennsylvania. From his regular reading of the medical journals, the doctor knew that RPR had a powerful new drug for Gram-positive infections wending its way through the FDA's approval process. Within hours, a precious package of what RPR was calling "Synercid" arrived at the hospital, sanctioned for Teresa on what the FDA called a compassionate-use basis—an experimental drug, not yet approved for the market, allowed in individual cases where no commercially available drug worked. Almost as soon as the drug began coursing through her veins, Teresa showed improvement, enough to stop throwing up, emerge from a coma, and start recognizing visitors. In two days, as if back from the dead, she was up and walking. Aside from an odd propensity to cause pain as it entered the body, Synercid appeared to have no serious side effects. For Teresa, the venous irritability, as doctors called it, was more than bearable, given the alternatives.

Teresa would remain in the hospital for two more months. She would have to wear a helmet when she returned to school that September to protect the exposed gaps in her skull until her original bone fragments, frozen in the interim, could be reattached. But the worst was over, and Synercid had made all the difference. "If we didn't have that drug," her father said later, "she wouldn't be alive."

That Rhone-Poulenc Rorer had even initiated development of Synercid when it did was surprising in itself. By the mid-1980s, roughly half of big pharma, as the large drug companies were known, had cut back on antibiotic research or stopped it completely. Why spend hundreds of millions of dollars on a drug for *S. aureus* and other Gram positives, they had reasoned, when the market for antibiotics seemed saturated already, and when vancomycin, the drug of last resort, seemed as effective as the day it had appeared in 1958? *Market* was the key word here, for at big pharma, marketers, not scientists, decided which drugs the companies should pursue. Microbiologists in the pharmaceutical labs could see resistance mounting to one antibiotic after another, write learned papers on the subject, and wring their hands at medical conventions. Out in the field, though, and even in the hospitals, physicians rarely complained of the problem—the clinical effects were subtle as yet—and physicians were big pharma's customers. To the company marketers, the big play appeared to lie with new drugs for diseases of the brain, heart, and lungs, and quality-of-life enhancers for baldness, depression, erectile dysfunction, and the like. A consumer would take drugs for those needs in an ongoing way, perhaps for the rest of his life. Such medicines would be vastly more profitable than antibiotics, which either worked or failed on a short-term basis against a patient's bacterial infection. So the industry felt content to turn out antibiotic knockoffs: third-generation penicillins, fourth-generation cephalosporins. Analogues, that is, of one tried-and-true drug or another.

Dr. Daniel Bouanchaud, head of antibiotic research at RPR, was one of the few in drug development who could see the bigger picture. In the 1940s and 1950s, he knew, when drug companies had come out with the first wave of antibiotics to fight *S. aureus* and other Gram positives and pronounced the battle won, the Gram negatives had surged. The drug companies had struggled to catch up, eventually, by coming out with a new class of antibiotics called the aminoglycosides, among them gentamicin and tobramycin; new, expanded-spectrum penicillins; and broad-spectrum cephalosporins. The campaign against Gram negatives had even continued through the 1980s with two new families of drugs, the carbapenems and the quinolones. But to Bouanchaud, the pattern was clear. Soon the Gram positives would stage their next resurgence, having grown resistant to all the old drugs used against them. And what, besides vancomycin, would be there to fight the worst of them?

In one sense, Synercid was anything but new. For nearly three decades, RPR had sold an oral antibiotic in France and Belgium called pristinamycin that treated a wide range of Gram-positive infections, including *S. aureus,* without provoking any significant resistance or side effects. Its mechanism of action was impressive: instead of binding to some element of the cell wall, it actually entered the bacterium and locked on to the cell's ribosomes to block the formation of proteins the cell needed to replicate. It was a natural drug that came from the fermentation of an Argentinian fungus called *Streptomyces pristinaspiralis.* Unfortunately, nature made it a complex and unpredictable mixture of nearly twenty molecules, each batch a bit different from the last. ("Chicken soup," one RPR scientist dubbed it.) France was willing to license it at the time because it worked so well, but no effort had been made to get it approved in the United States, because the FDA frowned on any drug that had more than two active molecules: too complicated and, the FDA feared, too risky. By the mid-1980s, however, new technology enabled Rhone-Poulenc Rorer to purify the drug—to fish out the best of those molecules, synthesize them, and then manufacture them together in an entirely predictable way.

Purifying pristinamycin took five years. Along the way, the chemists pored over other streptogramins—the drug class of which pristinamycin was a member—and found more promising molecules. They found that when two in particular were put together, they increased each one's potency by sixteen times. The chemists called one of the molecules quinupristin, the other dalfopristin. Like the molecules of pristinamycin, they attacked protein synthesis in the ribosomes of bacterial cells, but they worked in tandem: one hit the ribosomes at an early stage of protein synthesis, the other at a later stage. Against some organisms, the pair was bacteriocidal, against others it was just bacteriostatic, but either way the synergy—which eventually inspired the drug's name—was like a one-two punch.

For Bouanchaud's team, isolating the two molecules of pristinamycin that worked so well together raised an immediate problem. Both molecules, as it happened, were heavy—and fragile. As a result, they weren't soluble. That hadn't mattered before, because pristina-mycin was an oral drug. It mattered now, because a new drug for se-vere, systemic infections caused by Gram positives like *S. aureus*—the kind of infections vancomycin worked against—would have to be given intravenously, or so they thought, to get right to work. Resolutely, Bouanchaud's chemists spent the next *three* years making a soluble form of each molecule and testing the two together to be sure they didn't lose their synergistic punch. At the same time, they kept playing with the ratio of each drug to the other. If they used 10 percent quin-upristin and 90 percent dalfopristin, was the combination more pow-erful than with the ratios reversed? How about 80/20? It was dreary, frustrating work that often seemed doomed to fail. Two or three times a year, Bouanchaud had to plead with his superiors to keep the pro-gram alive. "We need money, and we need time," he would say, "but if we succeed, we'll have an amazing drug on the market just as *S. aureus* becomes the number-one enemy again."

Eventually, Bouanchaud's team did find the ideal ratio—30 per-cent quinupristin, 70 percent dalfopristin—but their superiors re-mained skeptical, and not just because the process was taking time and money. The very idea of a drug based on two molecules working synergistically made them nervous. Pristinamycin's many molecules worked that way, too, but having about twenty different ones to draw on increased the drug's chances of working. Asking two molecules— two drugs, in effect—to work in tandem posed a greater challenge. The combination might be unpredictable, and possibly unsafe. Even if an ideal ratio was found and both molecules had been rendered sol-uble, they would still have to be synthesized—an artificial, manufac-turable copy of them made—so that the final product had an ab-solutely consistent mode of action.

With his own company expressing doubts, Bouanchaud went to the U.S. Food and Drug Administration and pitched his story. The com-missioners were interested, but wary. They told Bouanchaud that RPR would have to conduct a set of phase one human trials—tests on healthy volunteers, to be sure the drug didn't do more harm than good—for each of the two drugs, followed by an unprecedented third phase one trial with the drugs together. Only then could RPR go on to phase two trials, on a select number of patients who were actually sick with the condition the drug would treat, and phase three trials, on a far wider group of sick patients. Bouanchaud relayed the FDA's verdict to his superiors. Everyone knew what that meant. It meant more time and money. Bouanchaud held his breath. He felt the most important contribution of his career was on the line. Grudgingly, RPR agreed to the extra phase one trials. With that, Bouanchaud's role began to taper off.

Now Synercid's fate fell to Sylvie Etienne and François Bompart, two French doctors in their early thirties. The two had just joined RPR—the year was 1991—to oversee the clinical trials of all new antibiotics, of which Synercid was just one. It was a job that had come to be as important as drug discovery itself. The cost of getting a new drug through the gamut of three-phase FDA trials to market had risen to $200 to $500 million, an industry average that included the costs of developing the losers as well as the winners. Synercid, it would be said by one scientist who worked closely on it, would cost Rhone-Poulenc Rorer more than $400 million. Etienne and Bompart had to be sure the trials occurred as quickly as possible. And if any were tainted by mistakes—faulty concoctions of the medicine, mismeasured doses, inaccurate results—the work of a decade could be derailed.

Etienne and Bompart's team conducted phase one trials for RP 59500, as the drug was still called, at RPR's Antony facility in France, near Paris, then in Collegeville, Pennsylvania. At the company's U.S. campus, several buildings of brown and white brick set a sternly corporate tone, punctuated by tall white poles with French and American flags. But some evidence remained of the forty-acre property's previous incarnation as a working farm: a barn and silo where employee training sessions were held, a lake, a cornfield, even a herd of grazing sheep. Here, Etienne and Bompart set about recruiting a group of doctors to administer Synercid to healthy volunteers. Lining up the doctors was easy: they had all heard about the idea of developing an injectable pristinamycin and were excited about it. For patients who volunteered, the only rewards were free checkups and the satisfaction of helping advance medical science. After a first IV infusion of Synercid, those incentives lost much of their allure. The damned drug *hurt*. After five days of twice-a-day dosage, even the most dedicated volunteers refused to submit to more.

Bompart's team never did figure out which of the two ingredients, quinupristin or dalfopristin, caused venous irritability, as they called it, when the two were used together. They did find, eventually, that they could dilute the drug in twice as much liquid—either a glucose solution or water then diluted by glucose—and reduce the irritability without having the drug lose any of its punch. Also, the pain was greatest when the drug was administered to peripheral veins, because its molecules were so large. If they administered it via a central line, a device allowing the drug to reach a large and deep vein directly, it hurt a little less, but it also diffused faster through the bloodstream. It was a tradeoff.

In phase one trials, Etienne and Bompart were less worried about venous irritability than they were about a potential showstopper: liver toxicity. Doctors administering the drug to healthy patients saw a disturbing increase in liver enzymes, a sign that the liver was having to work hard to flush out the drug's toxicity. By early 1993, RPR's senior management in France were nervous enough about the results to talk seriously of scuttling the whole program. Convinced that Synercid could beat the problem and emerge as a worthy alternative to vancomycin, Bompart wrote an impassioned memo to his superiors, outlining the case for keeping the drug in development, at least into phase two. The liver toxicity could be ameliorated, he argued, because it appeared only when the drug was administered at high doses. If it was given at lower dose levels and over a much larger patient population, Etienne and Bompart were confident the toxicity levels would go way down. After considerable debate, the drug was spared. Fortunately, Etienne and Bompart were right: lower doses in phase two trials still knocked out infections but without any noticeable uptick in liver enzymes.

No sooner had Etienne and Bompart's team survived that threat than they were forced to field another, this one in the form of a critique from one of the most distinguished medical researchers in the field, Dr. Robert Moellering. As physician in chief and chairman of the department of medicine at Beth Israel/Deaconess Medical Center in Boston, Moellering had enormous clout, especially in his specialties of vancomycin resistance and enterococci. He was one of the gods, like Courvalin and Hiramatsu, to whom big pharma sent all their new drugs for testing as soon as possible. When Moellering or another of the gods turned thumbs up, the new drug's success was all but assured. Thumbs down, and the drug almost as surely went back to the lab or got scrapped for good.

Synercid, Moellering conceded in his report, appeared to be bacteriocidal against many strains of *S. pneumo* and *S. aureus*, including methicillin-resistant *S. aureus*. But when used in vitro against 151 strains of vancomycin-resistant and intermediately resistant *Enterococcus faecium* and *Enterococcus faecalis,* the results were mixed.[1] Against Synercid, *E. faecalis* was pluckier. With more than a third of the VRE *faecium* isolates that Moellering tested, Synercid became effective at an MIC of 4 mg/L or higher—not a great result for the drug. Against VRE *faecalis,* half of the isolates were not inhibited by Synercid until it reached an MIC of 8 mg/L or greater. Ideally, a drug was bacteriocidal at 1 or 2 mg/L. And while the drug *inhibited* a substantial percentage of both kinds of VRE, it was not bacterio*cidal* against either. A bacteriostatic drug—one that inhibited but didn't kill—could still be useful against many kinds of infections. But, as Moellering observed dryly, a drug that failed to kill its target bacteria would be of no use with hard-to-reach heart valve infections. Nor, he might have added, would it be of much use with patients whose immune systems were too compromised to kick in and help the drug fight the bug.

RPR's scientists felt blindsided. From the start, Synercid had been

---

[1]The two species were slightly different from each other: *E. faecalis* was the more aggressive of the pair, but *E. faecium* caused a wider range of infections and seemed the hardier bug, especially when exposed to heat, which enabled it to survive better in undercooked foods.

envisioned as a drug to fight resistant *S. aureus* infections, as well as another growing resistance threat, *S. pneumo*. Since the drug's inception, VRE had begun to crop up in U.S. hospitals, but its incidence had remained spotty enough that Glenn Morris, in the VRE epicenter of Baltimore, was just starting to see how widespread it would be. And in Europe, in 1993, it remained a lab curiosity. Why judge the drug by a bug that seemed so insignificant—and one against which its creators had not intended it to be used? Besides, RPR's scientists muttered, Moellering had probably used the hardiest isolates he could put his hands on, from the most severely immunocompromised patients. A lab as prestigious as Moellering's, Bompart decided with Gallic humor, would no doubt get the toughest bugs.

The team was still stewing over Moellering's indictment one day in June 1993 when Bompart, by now in charge of the team, got a call from New York Hospital Queens. Dr. James Rahal, the hospital's head of infectious diseases, had a desperately ill patient on his hands. The patient, a forty-six-year-old woman, had a history of heart disease and had recently undergone several procedures to repair and replace damaged heart valves. After surgery and weeks of hospitalization, she had contracted a whole host of infections: *Pseudomonas aeruginosa, Acinetobacter baumanii,* MRSA, and, along with all of those, VRE. Since then, Rahal had tried everything: chloramphenicol, ciprofloxacin, doxycycline, and ampicillin. The bacteria resisted them all, and now the woman was dying. Rahal had called his lab director to ask if by chance there was any other antibiotic he'd heard of that might help. The lab director, it turned out, was the one person at New York Hospital who had heard of RP 59500, through a friend who worked at RPR. The woman was about to die, Rahal's lab director stressed in his call to Bompart. Could he get some RP 59500 right away?

Bompart felt guilty and embarrassed. He told the lab director about Moellering's recent indictment of RP 59500. The report had so depressed the RPR scientists that they hadn't yet even thought to test their drug in vitro on VRE *faecalis* or *faecium* themselves. Now Bompart worried that it might not work with the New York patient. Then Bompart would have to account to his superiors for a first, unexpected, compassionate-use in vivo test that ended with a patient's death from various resistant infections—he could take his pick. That would doom the program, he was sure of it. But how could he keep the drug from a patient whose life might be saved by it? "I don't think it'll work," Bompart said with a sigh. "But of course you have to try. Just try it in vitro first, okay?"

Within hours, a precious dose of RP 59500 was dropped onto an isolate of the woman's infected blood. To Rahal's delight, it did inhibit the VRE *faecium*. Thrilled, Bompart sent more to be given directly to the patient after getting FDA approval for compassionate use of the drug. Almost immediately, her VRE *faecium* and MRSA began to diminish. For twenty-five days, Rahal kept her on a steady diet of RP 59500, along with other antibiotics to eradicate the Gram-negative infections. The patient did experience liver toxicity, but only mildly, and had no other adverse effects. At the end of that time, Rahal pronounced her cured.

Other compassionate-use cases followed—a forty-nine-year-old woman in Arlington, Virginia, with a heart valve VRE *faecium* similar to the New York patient's; a seven-month-old baby girl in Long Island, New York, with VRE *faecium*-infected spinal fluid—and, in both, RP 59500 worked with no serious side effects. In Collegeville, RPR scientists eventually conducted their own in vivo tests of the drug with VRE *faecium* isolated from sick patients and found that the susceptibility rate of VRE *faecium* was more like 95 percent than the 65 percent overall susceptibility that Moellering had encountered. Either Moellering did have unusually resistant VRE bugs in his lab, as RPR scientists had imagined, or the drug did better in vivo than in vitro.

These successes did more than save the program. With VRE *faecium*, they also got the drug over the catch-22, as Bompart put it, of gaining FDA approval without a "comparator." By the FDA's rules, every new drug in phase two tests with sick patients had to be tested against a comparable drug. There could be, in other words, no double-blind tests with placebos, not when people's lives were at stake; every patient would receive either the new drug or the comparator. The catch-22 was that with VRE, there *was* no comparator for Synercid: no other drug worked consistently against VRE. But as the list of compassionate-use cases grew to more than fifty, that no longer mattered.

Still, there were other hurdles. By FDA rules, patients in a trial had to have no other infections than the one being treated, or the results would be muddied. But the likeliest candidates for MRSA were intensive care unit patients in such poor health that they had multiple infections. To Bompart's great frustration, the team never did find enough pure MRSA patients to conduct MRSA phase two trials. And because vancomycin ostensibly worked with those patients, the FDA saw no reason to cut Synercid any slack. The same problem of multiple infections applied with VRE *faecium* patients, but the FDA was more tolerant, because with VRE, Synercid was a drug of no choice: vancomycin had failed with those patients. As a result, Synercid emerged from phase two with FDA approval for VRE *faecium* and, as it turned out, methicillin-susceptible *S. aureus,* as well as *S. pneumo.* But not approval for MRSA.

Bompart stayed in RPR's clinical testing division long enough to see Synercid through phase three trials, and he was joined at that time by an American physician, George Talbot, who oversaw the VRE *faecium* studies. Now patients from around the world were recruited, under the auspices of doctors like Moellering, to try Synercid for VRE *faecium,* methicillin-susceptible *S. aureus,* and *S. pneumo.* Some 2,000 patients in all were given Synercid. In 52 percent of those cases, the drug killed or inhibited severe infections. So it worked only half the time, and even when it did many of the "cured" patients were too sick to survive. But these were infections that had resisted vancomycin — the worst infections one could find, likely to kill most of the patients afflicted with them. The drug did well with *S. aureus* and VRE *faecium,* less well with community *S. pneumo,* due to the wide range of strains roaming outside hospitals. All things considered, it seemed a great success. Even Robert Moellering found reason to praise it: in a second study of Synercid, he determined that of 396 patients with VRE *faecium,* 65.8 percent were cured of their infections. The drug, he added, was generally well tolerated.

In the long gamut of drug approval, RPR still needed to submit cartons of trial results and supporting information about Synercid to win
FDA registration, a laborious task that fell to a cheerful, resolute microbiologist named Harriette Nadler. It also had to *produce* the drug.
After all the work and risk-taking, the years and careers and stag-
gering sums of money, Synercid nearly died at that stage, too. In
September 1997, after the whole manufacturing process of the drug
had been mapped out in an assembly space in Kankakee, Illinois, the
FDA made a routine inspection of the plant and declared various as-
pects of the plant not in compliance with FDA production standards.
Overnight, the whole place was shut down.

One month's delay became two, then ten. The drug was still available on a compassionate-use basis, so lives were being saved. But a
near-perfect market window was being squandered. So, for RPR,
were potential profits on a $400 million investment. For the company,
there was another reason to agonize over the production delays.
Singular as it may have seemed at its inception, Synercid now had a rival. And thanks to the shuttered plant in Kankakee, that rival might
just beat it to market.

To the rattled researchers of RPR, linezolid was a latecomer, and a
vexing one at that. For Chuck Ford and his colleagues at Pharmacia
& Upjohn, it was the outcome of more than a decade's work and faith
that had started, for Ford, on a feverish day in New York City in 1987.

Ford, a Michigan-born-and-raised microbiologist, had come with
his wife to New York partly on business, to attend a medical conference, and partly for fun. Almost immediately, both had succumbed to
keen gastrointestinal distress—punishment, or so it felt, for being local yokels braving the big city. At the medical conference, Ford
winced when his buddy Steve Brickner, an Upjohn chemist, insisted
on dragging him over to see the poster for a new drug announced by
DuPont. The last thing Ford wanted was to take any unnecessary
steps. "Chuck, you have *got* to see these drugs," Brickner declared.

The two stood by the poster, Ford leaning against the wall for support. *"Man,"* he said at last. "This stuff is too good to be true."

In the long-established culture of medical conventions, drug companies put up fact-filled posters to announce new drugs in development. Typically, the new prospect was such a specific variation of an existing drug that patents could protect it. But DuPont 105 and 721 appeared to be the first drugs of a whole new class, one DuPont was calling oxazolidinones (oxys for short) and, as the Upjohn scientists knew, you couldn't patent a whole class.

If DuPont's claims could be believed, oxazolidinones were active against all Gram positives—*Staphylococcus,* enterococcus, *Streptococcus,* and more. They were different from Synercid and the streptogramins in a profound, fundamental way. Synercid had come from nature—that fermented Argentine fungus. Eventually its molecules had been synthesized, but only to replicate what nature had provided. The oxazolidinones, by contrast, were new chemicals created by humans. They had come out of thin air. No bacterial pathogens, as a result, had ever seen them or anything like them. Presumably, the bacteria would be unable to defend themselves against them, for the short or maybe even the long term.

Not only that, Ford marveled: they were being proposed as an oral drug. That would give DuPont's new class a huge edge over vancomycin if the drugs could actually get past three-phase trials to market, because they could be used in the hospital *and* the community. They didn't have to be administered, as vancomycin did, intravenously by central line, restricting their use to the hospital and nurse-assisted home care. When Ford got to the section on adverse effects, he whistled: there didn't appear to *be* any. "Hey, Steve," Ford said, his gastrointestinal distress forgotten, "let's go home and make some oxys."

Back in Kalamazoo, Ford, Brickner, and a third colleague, Gary Zurenko, began to tinker with oxys as a private project, or skunk works. It was a secret from all but the scientists' senior management, for in the drug business, as in any very competitive field, rivals had a way of learning about each other's new programs and jumping in to catch up. Upjohn encouraged such secret fiddling: 20 percent of each scientist's time was allocated for it. But no troika of scientists had ever pushed an entirely new drug up to management in the way that Ford and his colleagues began to do. They just had faith, they decided later. The drug, they felt, was that good, and the need so obvious. Over the next five years as they played with the chemical structure of oxys, VRE remained a distant threat. But MRSA was a huge and growing one, treatable with only one drug, vancomycin, and even that drug seemed threatened. Cavalier as the industry might be toward Gram positives, Ford saw a looming market opportunity, of which he was, perhaps, more mindful than many drug company microbiologists. Back when Ford had joined Upjohn, his father, a businessman, had given him good advice. "Don't just do the science," his father had said. "Be smart—learn the business." Ford had listened. He'd hung out with Upjohn's marketing and sales people and found out everything he could about the business of antibiotics: not just how they were made but how they were sold to doctors and hospitals. Now as a "Scientist Six" in Upjohn's hierarchy, a top researcher in the corporate flowchart, Ford knew he had a potential blockbuster.

In 1990, Ford and his team made a formal pitch for a full-fledged oxy program to Upjohn's senior management. First they explained how the drugs actually worked. Instead of binding to some part of the cell wall or stopping the bacteria from using essential proteins, oxys disabled the protein-making process at its source. They inhibited RNA from conveying genetic instructions from DNA to the ribosomes—the protein factories of the cells. And they kept the ribosomes themselves from assembling proteins. (The target was similar to Synercid's: the French drug attacked the ribosome as oxys did but did not target RNA). The drug's oral form—a consequence of the class's relatively small molecules—was a huge plus, and so far, Ford went on, no pathogen had been able to create resistance to it in vitro. Oxys disabled so many genes, he explained, that the bug would need to perform a miraculously elaborate exercise in genetics to compensate for them. True, Courvalin had shown that enterococci could orchestrate such a process against vancomycin. But it had taken the bugs thirty years! By then, Upjohn could have a whole mess of other oxys on the market. Look at how many generations of penicillin the industry had cranked out, he said, even with ever rising resistance. Upjohn could do the same with oxys.

Ford's presentation had to be "warts and all," so he duly ticked off the one risk factor the team had seen. A lot of the oxys tested so far had an occasional tendency in laboratory animals to suppress stem cell production in bone marrow. That could lead, in turn, to suppression of white blood cells and platelets, essential parts of the immune system. Ford was confident, though, that his team could isolate an oxy that didn't do that. Overall, the drug's toxicity profile was gorgeous, as Ford put it. No drug was perfect, but this one seemed pretty close.

Competition was the other risk factor, though this was hardly unexpected. If Upjohn had to race against a rival, DuPont was a manageable one. Only recently had the company even entered the antibiotics field; it had a lot to learn, Ford felt. At the time, he had no idea that two other companies, Bayer and AstraZenica, had launched oxy programs of their own. Nor did he know that Rhone-Poulenc Rorer was working hard on Synercid. Knowing about these rivals would have done nothing to dampen his enthusiasm, however. No matter what drug you started with, no matter how promising it seemed, bringing it to market was a monstrous crapshoot. You had to take the best drug you could find, with the fewest downsides, and hope for the best.

Ford's superiors were impressed enough to sanction the program and give his team the use of five full-time labs. Then the real slogging began. Dozens, then hundreds, of oxys were synthesized, each next one the slightest bit different in its chemical structure from all that preceded it. All turned out to be less effective, or more toxic, than hoped. Colleagues in other departments began rolling their eyes. "How many of these are you going to make," they'd say to Ford, "before you realize this is a blind alley?" At times, even Brickner grew discouraged. He'd get some small incremental improvement with the next chemical variation, but then nothing better for months— months of tedious testing. He'd turn to Ford with a wry look and ask, "Do you think we're hitting the wall?"

While the program was still at the in vitro stage, Ford heard a shocking piece of news: DuPont had dropped out of the oxy business. Rumor had it that the company's best prospect, DuPont 721, had proven toxic in every animal species on which it was tested. Grimly, Ford and his team took their three best prospects and embarked on animal tests of their own. One of the three, they found, was absolutely, unquestionably nontoxic. Later, when Upjohn put up its own poster at one of the medical conferences, announcing that it was taking an oxy into trials, the DuPont scientists would stare at it, stunned. "How in the hell did you guys think you could make a nontoxic oxy?" one of them would ask Ford. The Upjohn scientist wasn't sure himself. "I guess we just believed we could."

By 1994, the team members felt ready to put a leading compound into phase one trials. Maddeningly, they had *two* leading contenders— two that appeared, in the test tube and in animal models, to have identical effects. There would be no way to tell the two apart until the team saw how they acted in humans. With management's reluctant blessing, the team decided to put both through phase one trials, at twice the cost. That was, to say the least, an unusual decision, but Ford and his colleagues felt they couldn't risk pinning their hopes on one, having it bomb out, and feeling haunted for the rest of their lives that they might have left a winner behind. One of the two, dubbed eperezolid, was put through phase one trials with healthy volunteers in Kalamazoo. The other one, linezolid, was put through phase one by Japanese colleagues. Soon the verdict emerged: linezolid worked longer before the body dispelled it, which also meant the drug could be administered to a patient less often in the course of a day.

Ford, ever mindful of the business side, wanted a clinical doctor's reactions to the drug. One of Dr. Don Batts's findings was that the drug had a curious propensity to be excreted, in high doses, through the skin. One patient swore he could smell the drug on his hands. Batts thought not. The two made a wager. First the patient received a placebo: no smell. Then linezolid. "Yup, that's it," the patient declared. Ford was thrilled. He'd keenly hoped that linezolid would zap skin and soft tissue infections: *S. aureus, S. epidermidis,* and many of the streptococci. In order to do that, however, it had to disseminate well, not only through the circulatory system but also out into the musculature and other soft tissues. Because it was the first drug of an entirely new class, Ford had no sure sense of how well it would do that in a human being. If that guy was smelling the drug on his skin, it had traveled, all right—as far as Ford could have hoped.

Ford and his team had been delighted that oxys worked as oral drugs, and they had given no thought to developing an intravenous form of the drug. To them, that seemed like going backward. But Batts insisted that hospital doctors, especially those in intensive care units, would never take seriously a drug that came only in oral form. Upjohn might prove in tests that oral linezolid worked as expeditiously as any IV drug, but the doctors simply wouldn't believe it. The scientists had only to imagine a scene from the hit medical series *ER* to see what Batts meant. In would come a desperately ill patient on a gurney to the ICU. Could they imagine the doctor played by George Clooney turning to his colleagues and saying, "Give that patient a pill"? Pills were convenient, all right, but they would never seem as fast-acting as an IV. And in the ICU, somehow, they just weren't macho enough to impress the docs whose every decision could determine life or death.

Because oxys were small-molecule drugs, the team had relatively little trouble adapting them to IV, as Rhone-Poulenc Rorer's chemists had had with Synercid. Indeed, oxys could be administered through any peripheral vein, cause no pain, and, as the team had just seen, disseminate quickly through the body. That gave doctors an advantage that might prove helpful in certain cases—without any sacrifice of medical macho.

By the onset of phase two trials, another possible problem had emerged. Oxys appeared to inhibit monoamine oxidase, an enzyme lining the gastrointestinal tract that breaks down certain hard-to-digest foods like salami, aged cheese and wine, brewed beverages, and soy sauce. Conceivably, if patients on oxys ate those foods, their monoamine oxidase would be too inhibited to convert those foods into usable nutrients. That, in turn, would raise blood pressure and heartbeat. Upjohn sent out stern advisories to doctors conducting the phase two trials, asking them to tell patients not to consume those foodstuffs.

The good news from phase two trials was very good indeed. In twenty-five separate studies with sick patients, linezolid proved effective against all *S. aureus* and *S. pneumo* symptoms. It handled all enterococci, too—not just *E. faecium,* as Synercid had done. The drug was essentially inactive against Gram-negative bacteria, but so what? It couldn't be expected to obliterate all the bad bugs. None of the phase two patients experienced any inhibition of their monoamine oxidase levels, even though, as the team learned later, the participating doctors had largely neglected to pass along Upjohn's food warnings to their patients. As for the more serious concern about bone marrow suppression, a few patients experienced a slight suppression of stem cell production when given multiple doses of the drug. But as soon as they were taken off the drug, Don Batts reported to his superiors, stem cell production bounced back up. At worst, this side effect appeared to be rare and reversible.

Linezolid had passed phase two with flying colors. Yet no sooner were the results announced than Ford began to hear a low drone of skeptical muttering from doctors in the field. Linezolid was bacteriocidal for *S. pneumo,* they observed, but only bacterio*static* against *S. aureus* and *E. faecium* and *E. faecalis.* A drug that only inhibited those bacteria might work sometimes, they said. But what about in cases of endocarditis—heart infection—or infections from bone marrow suppression? Any drug that failed to act quickly against those urgent, hard-to-reach infections would be all but useless.

Ford did a little homework and learned that the definitions of *cidal* and *static* had blurred since the 1950s, when they originated. The word *static* had not been meant to indicate that a drug failed to kill its target bacteria, only that it failed to kill a specific percentage of them within a certain period of time. Yet doctors had come to feel that any drug called bacteriostatic was a drug that only inhibited the bacteria from replicating. Ford looked hard at linezolid's action in the phase two trials. Linezolid remained static against staphylococci and enterococci, not cidal, because more than .1 percent of the bugs remained alive twenty-four to forty-eight hours after the drug was administered. Yet far sooner than that, patients showed dramatic signs of improvement. One doctor, Judy Stone of the Baltimore VA medical center where Glenn Morris worked, found that among patients with severe staph skin infections—patients whose backs or chests were utterly covered with excruciating infections—oral linezolid was incredibly fast-acting. In twenty-four hours, the drug reduced infections by 50 percent. In forty-eight hours, they were gone. Yet, presumably some of the bacteria remained alive. Finally a colleague of Ford's dispelled the mystery. Even minute quantities of linezolid, he found, almost immediately shut off the bugs' production of toxins. The toxins were what caused the infections. Who cared if some of the bugs were still alive after forty-eight hours, Ford began to preach to doctors, if they were defanged, and the infections they'd caused had healed? The doctors began to listen.

Another threat seemed to loom that fall of 1995—the merger between Upjohn and Sweden's Pharmacia AB. Continuing a trend that would see the industry consolidate into a handful of international giants by the turn of the century, the marriage of Upjohn and Pharmacia endangered every new drug program at either company that carried any degree of risk. To the great relief of Ford and his team, however, the Swedes were way ahead of the U.S. drug industry in appreciating the dire implications of antibiotic resistance. They needed no convincing that Gram positives had staged a major resurgence and constituted an imminent global crisis. They just wanted to know how soon Upjohn could get linezolid to market.

By now, Ford was well aware that Rhone-Poulenc Rorer had Synercid in the works: Robert Moellering's lukewarm appraisal of it had appeared in a medical journal in March 1993. Synercid was targeting the same Gram-positive bacteria, but at least it wasn't an oxy.

Ford was more worried about AstraZenica, whose own oxy was clearly moving through development.[2] Every time Ford's team submitted new information for another patent on some aspect of oxazolidinones, AstraZenica would apply within months for almost identical patent protection. Happily for Ford, P&U was just far enough ahead to get most of what it applied for. He could only imagine how vexed his counterparts at AstraZenica must be.

Both from phase three trials and from compassionate use, the number of sick patients exposed to linezolid now increased dramatically. And so did the dramatic results. One remarkable story came from Orlando, Florida, in February 2000, when a teacher in her mid-thirties—a nondrinker and a nonsmoker, healthy and fit until then—got leukemia, which ravaged her immune system and rendered her susceptible to a VRE infection. No problem, her doctors said brightly: they had a new drug called Synercid. Unfortunately, Synercid failed. Fortunately, the doctors had linezolid as a new last resort. In five days, the teacher tested negative for VRE. She still had leukemia to contend with, but at least she had a fighting chance to beat it.

The overall figures were just as impressive. In one typical trial study of 200 patients, linezolid cured 95 percent of skin infections, mostly *S. aureus,* and 96 percent of bacterial pneumonia cases (including *S. pneumo*). In an MRSA trial of 460 patients where linezolid was compared to vancomycin, the cure rate for linezolid was 77 percent versus 74 percent for vancomycin. In a study of forty-four patients with vancomycin-resistant enterococci, there were, of course, no comparator drugs. But 89 percent of those patients were cured with a maximum dose of linezolid.

One of the many doctors around the world who conducted these trials was Robert Moellering, who grudgingly admitted that linezolid "worked pretty well," and that it represented the first entirely new class of drugs "to get to this point of development in thirty-five years." "All of the data so far," Moellering declared, "suggests that

[2]Bayer, at least, had dropped out of the oxy race.

this is a drug for which it will be difficult for bacteria to become resistant.

"But," he added, "we thought that was true for vancomycin, too."

In the end, Synercid did beat linezolid to market, by a little more than six months. In September 1999, formal approval from the FDA came for Synercid as the first alternative in thirty years to the antibiotic of last resort, vancomycin. It was approved for VRE *faecium* infections and came as a godsend: by then, according to the CDC, more than 15,000 Americans were becoming sick with *Enterococcus faecium* each year. In half of those cases, the bugs were resistant to vancomycin. Unfortunately, Synercid was still not approved for MRSA or *S. pneumo* or VRE *faecalis*. The FDA's antibiotics chief, Dr. Sandra Kweder, suggested that the narrow range of indications approved for its label might help protect Synercid as a drug of last resort: the less widely it was used, the later resistance might arise to it. But RPR could hardly be pleased. The narrower the range of indications, the fewer patients would be eligible for the drug in which it had invested nearly half a billion dollars.

Linezolid, rechristened Zyvox for its market launch, won formal FDA approval in April 2000. The agency approved the drug for a slightly wider range of indications than Synercid: not only VRE *faecium* but VRE *faecalis*, hospital as well as community *S. pneumo*, and methicillin-susceptible *S. aureus*. But in the MRSA trial, the FDA adjusted the cure rate for Zyvox down from 77 percent to 56 percent after a case-by-case examination in which several criteria for cure were applied: if a case failed to meet even one of those, regardless of whether a doctor had proclaimed a clinical cure, it was disqualified. As a result, linezolid, like Synercid, was denied an indication for MRSA. Whether or not doctors would go "off label" with the drugs, prescribing them for MRSA anyway because drug company salesmen showed them how well their products had worked on MRSA in clinical trials, no one could say. The fact was that while both Rhone-Poulenc Rorer and Pharmacia & Upjohn could take pride in getting drugs to market, the two had spent a collective billion dollars in research and development and still had not come up with a drug with an approved indication for MRSA—the most virulent bacterial infection of them all and the one that had just devised a way of becoming partially resistant to vancomycin, the one drug that still worked against it.

Because U.S. patent law defined the patentable period for a new drug to be seventeen years, a drug company that had spent a decade bringing a drug to market had just seven years to make its money back before the onslaught of generic products from rival companies. With an eye on that clock, both Rhone-Poulenc Rorer and Pharmacia & Upjohn began spending all the money they could to promote their new drugs—to establish the brand, as the advertising world put it, and saturate the market. Sales representatives gave out generous supplies of samples to every relevant doctor in America. At medical conventions, doctors received handsome products emblazoned with the new drug's name: umbrellas, luggage, warmup jackets, and more. (By one estimate, every major U.S. drug company spent on each relevant U.S. doctor an average of *$5,000 per year* in promotional materials, including free drugs.) With most kinds of new drugs, a huge marketing campaign might do no medical harm. With antibiotics, it was, perversely, the worst possible way to go. The more a new antibiotic was used, the more quickly bacteria managed, by mutation or importing a gene, to develop a resistance mechanism to it.

As the drugs hit the market, one drawback was obvious: both were appallingly expensive. Penicillin had cost pennies a day. Methicillin, no longer in common use because of increased kidney toxicity, as well as rising resistance, had cost a few dollars a day, vancomycin about $14 a day in its maturity. Zyvox, by contrast, cost about $140 a day, while Synercid cost between $180 and $250 a day. Patients without medical insurance in Western countries could ill afford such drugs. For patients in developing nations, they were completely out of reach.

Perhaps the other cause for concern should have been no surprise either, though it always seemed to be with each next so-called miracle antibiotic. Within a year of each drug's launch, the first reports of resistance came in.

Pharmacia & Upjohn had dutifully reported two instances of VRE resistance to linezolid back in 1999, both in compassionate-use cases, among hundreds of cases. In both, patients had been on multiple drugs for various infections and had had in-dwelling medical devices, such as heart pumps, on whose surfaces the organisms had adhered. These were the hardest kinds of infections to reach and clean out, and they tended to grow in extremely sick patients. By the drug's launch, fifteen such cases had been reported. All involved extreme conditions and seemed to have little or no bearing on the drug's real prospects. But a year later Dr. John Quinn of the University of Illinois's College of Medicine in Chicago reported more disturbing results.

Quinn used Zyvox on five patients with vancomycin-resistant *E. faecium* infections. One patient had been infected as a consequence of heart surgery. The others were organ transplant recipients. Initially, Quinn found, Zyvox was effective in all patients. But in each case after three or four weeks the bacteria began to develop resistance to the drug. Admittedly, the patients were all very sick to begin with. But not all were old: one of the four was a twenty-eight-year-old man in the hospital for a liver transplant, another a twenty-four-year-old man with leukemia who had to undergo bone marrow transplantation. In three cases, when Zyvox began to fail and the patients appeared in danger of dying, Quinn switched to Synercid. Nevertheless, two of those patients died of uncertain causes; the third was in intensive care at the time of Quinn's report.

Dr. Cameron Durrant, Pharmacia & Upjohn's vice president of infectious diseases, observed that enterococcus was the only bug to resist Zyvox out of some 20,000 strains of resistant bacteria tested to date. Even in those cases, he added, the drug appeared to fail only in patients with severely compromised immune systems or with implanted medical equipment such as catheters or heart pumps to which the bacteria could cling and survive despite the presence of the drug, or in patients receiving Zyvox for a prolonged period of time.

So far, Durrant added, 60,000 patients had received the drug, without a single report of Zyvox resistance.

In a follow-up study to Synercid trials, RPR had duly noted rare instances of resistance to the drug in various bugs. Resistance to Synercid occurred in 6 of 338 cases of VRE *faecium* and in four of these led to therapeutic failure. Though the drug had not been sanctioned for MRSA, it could be used on the bug in emergency-use situations — basically, when vancomycin failed. In those cases, it worked about 66 percent of the time: a flawed but still impressive record.

In Michigan, however, a more ominous study had been conducted by Marcus Zervos, M.D., of William Beaumont Hospital in Royal Oak. Zervos had tested 100 random Minnesota residents and found Synercid resistance in about 2 percent of them. That was a fairly high incidence at face value. The real surprise of Zervos's study, however, was its date. The study was conducted in 1997, *before Synercid had reached the market.* None of those 100 residents had ever been treated with the drug. Where had the resistance originated?

Zervos felt he knew. For nearly three decades, American farmers had used an antibiotic called virginiamycin as one of their "growth promoters" in animal feed. Virginiamycin, as it happened, was an analogue of Synercid.

What Zervos's results suggested was that, over time, enterococci and other bacteria in animals had learned to be resistant to virginiamycin. That resistance had transferred to the humans who cared for, processed, or ate those farm animals. No one had bothered to test humans for resistance to virginiamycin because virginiamycin wasn't used in human medicine. Even in the early 1970s, it had been found to be too toxic at therapeutic levels for human use. But if Zervos's hunch was correct, a lot of people were carrying around resistance to virginiamycin in their gastrointestinal flora. Because Synercid was an analogue, those flora would be perfectly capable of mustering resistance to it, too.

To Zervos, an all-too-plausible scenario presented itself. As Synercid began to be widely used, its bacterial targets would do what any resourceful bugs would: cast about for resistance mechanisms to repel it. And there would be virginiamycin resistance genes, floating right around in the gastrointestinal neighborhood, precisely engineered for the job. Clearly, this horizontal resistance could happen: that was why 2 percent of those human patients had proven to carry Synercid resistance.

If Zervos was right, Synercid was not merely threatened. It was doomed—by the same industry that had developed it.

7

A Deadly Threat in Livestock

Shy and unassuming as he seemed in person, Marcus Zervos nursed an intense passion about drug-resistant bacteria. He had, as he put it, devoted his life to the enterococcus. The decision had come in the 1980s when, as a fellow in infectious diseases, he had witnessed a violent hospital outbreak of enterococcus that was resistant to the antibiotic gentamicin and led to deaths from bloodstream infections. In an immunocompromised patient, the bug was no second-rate pathogen— it was *serious*. Yet so little about it was known. By the time VRE spiked in the early 1990s, Zervos had become an expert. Which was why, beginning in October 1995, he spent a whole year studying turkey feces.

The feces came from three flocks of turkeys, each consisting of about 30,000 birds, on two large rural farms in western Michigan. As per the farms' standard practice, the turkeys in Zervos's study received small daily doses of virginiamycin from nine and a half weeks of age until just before slaughter, at about forty-two weeks old. Zervos found that by the time of slaughter 100 percent of the *E. faecium* isolates cultured from roughly 200 animals were resistant to virginiamycin. More disturbing, all 100 percent were cross-resistant to Synercid.

Until Synercid actually reached the market and began bombarding VRE in human patients, no one would know if human VRE bacteria could snag Synercid resistance genes from animal enterococci traveling through the human gut. As one researcher put it, the human gut was still a dark and mysterious place to microbiologists. They could deduce that animal enterococci absorbed by a human appeared to pass resistance genes to the human's own enterococci by analyzing, to put it politely, the end product. But they couldn't be absolutely, scientifically sure that that particular bug had transferred resistance to that particular other bug. And they could only guess what might happen when new combinations of bacteria met in the gut.

Zervos's follow-up study, in which he found that 2 percent of 100 Michigan residents had Synercid-resistant enterococci in their stools, was so startling, given the imminent arrival of the drug, that researchers quickly did follow-ups to see if Synercid resistance could be confirmed in other locales. One of those researchers was Glenn Morris, whose Baltimore lab often worked with Zervos's. Morris had not been surprised by the Michigan results. In 1996, he and his researchers had gone to test fecal droppings at various state and county fairs, where their serious demeanor and meticulous gathering methods earned them a lot of bemused looks. Undaunted, they collected droppings from twenty-five chickens, twenty pigs, and twenty-five cows. The animals were from farms throughout the state. Resistance to several drugs was found, gentamicin prominent among them, which led the team to make the then quite controversial suggestion that multidrug-resistant enterococci did colonize poultry and pigs and that the resistance might pass to humans. Almost in passing, the researchers had also noted that three strains of Synercid-resistant *E. faecium* were found among the isolates—rather startling, given that Synercid was years away from reaching the market.

In a large, windowed office in one of the CDC's newer buildings in downtown Atlanta, Fred Angulo pondered Zervos's and Morris's findings with a sense of grim foreboding. As the CDC's top bureaucrat on animal-use antibiotics,[1] Angulo had more power to act on

---

[1] Specifically, Angulo was head of two programs, FoodNet and the National Antimicrobial Resistance Monitoring System, within the CDC's Foodborne and Diarrheal Diseases branch.

those findings than any research scientist or doctor, but often at times like this he felt his hands were tied. As an advisory body, the CDC could only recommend, never command. Often, the best the CDC could do was order follow-up studies of its own to lend the agency's stamp of approval to a lone investigator's results. And so, at Angulo's instigation, CDC investigators purchased chickens from twenty-six grocery stores in four states, and found that nearly 65 percent of them had Synercid-resistant *E. faecium*. These, too, were results obtained before Synercid reached the market.

In none of these studies, at least, were the enterococci in poultry resistant to vancomycin. They were not, in other words, VRE. Wolfgang Witte had found VRE in European chickens because Europe used avoparcin, and vancomycin was an analogue of avoparcin. No avoparcin was used in the United States, due to the FDA's ban on it from the start, so no VRE had materialized among U.S. animal enterococci. In the United States, as a result, the immediate threat to people seemed almost nil. Even if the poultry bugs did pass resistance genes for virginiamycin to people, that wouldn't matter, because virginiamycin wasn't an antibiotic for people. Neither was virginiamycin's analogue, Synercid—yet. But given how easily enterococci appeared to pass resistance genes to one another as a general matter and how easily VRE circulated in hospitals, how long would it be before Synercid-resistant VRE was ubiquitous? Before the new miracle drug was dead?

Angulo, a large, broad-shouldered man whose Colombian lineage lent him a brooding intensity that made him seem older than his thirty-nine years, could disseminate these findings through the CDC's journals. He could pass them on to the FDA's Center for Veterinary Medicine, which did have the power, within tight constraints, to regulate which antibiotics reached the animal market. But his greatest power was to alert the press and consumer groups to new data. It was a delicate balancing act, trying to maintain objectivity as a government bureaucrat while trying to help the country come to terms with a creeping crisis, but Angulo did it well. Like his CDC colleague Fred Tenover, he had several degrees—two Ph.D.s and a master's in his case, including one in veterinary science—and nearly fifteen years of experience in the CDC's food-borne diseases branch, which he had joined as an EIS officer. Like Tenover, too, he viewed his work as more than a job. It was a moral commitment. He remained scrupulously bound to the facts, but often he could see where those facts led. This latest revelation only underscored that. As a CDC official, Angulo would follow all the steps of due process in determining if a recommendation should be made to the FDA to have virginiamycin yanked from the market. Personally, though, he had no doubt that virginiamycin posed a real and immediate threat.

Personally, too, he had little hope that he or anyone else could do anything about it.

In a set of offices on G Street in Washington, D.C., rather grander than any at the CDC, Richard Carnevale often muttered Angulo's name in wry deprecation as he made the case for one animal-use antibiotic or another. Carnevale was about the same age as Angulo and had started his career in the same way, training to be a veterinarian, then joining the government. But then the two men's paths had diverged, and here was Carnevale, a couple career turns further, as vice president for regulatory, scientific, and international affairs at the Animal Health Institute, a title that meant he was a lobbyist for the trade group of big drug companies that made animal-use antibiotics and big food producers that used them. A trim, dark-haired man of modest height given to expensive suits, Carnevale was as good at his job as Angulo was at his own. If asked, the lobbyist would say that he worked to get vital new drugs into the hands of veterinarians, and that was true, as far as it went. His underlying task, though, was to sprinkle doubt and suspicion on each new study that raised concerns about animal-use antibiotics. His name popped up in virtually every newspaper report on the subject: the obligatory voice of the industry. To those who paid his salary, Carnevale was a trooper, doing battle with the obfuscating forces of the federal government and academia—the Angulos who wanted to ban good drugs on the basis of some flimsy study by a university doctor of four animals in Omaha. To the other side he was, as he put it himself, the Black Hat.

With virginiamycin, Carnevale had soothing logic, as usual, to proffer. In its quarter-century of use, he observed, the drug had caused no resistance problems in humans. Here, indeed, was a case of the industry bending over backward to comply with CDC concerns, using a drug that in all that time was an analogue to *no* human-use antibiotics. Now came Synercid, after extensive clinical trials. And still, in all those patients tested, hardly any resistance problems had been noted. "If virginiamycin had caused lots of resistance in people for the twenty-five years it was used," Carnevale observed, "you'd expect Synercid wouldn't work."

This was, as so many issues were in the murky realm of animal-use antibiotics, a matter of interpretation. No one could argue that virginiamycin had been used for a quarter-century without causing resistance problems in humans. And, as Carnevale observed, Synercid's clinical trials had not disclosed an immediate problem of cross-resistance. But why? Angulo believed that for animal-use antibiotics to cause resistance in humans you needed both the "seeds" and a force to "germinate" those seeds. The seeds were the animal-use antibiotics and the resistance they transferred to people. The germination was the prolonged use of a human antibiotic in the same class as the animal-use drug. It was true, Angulo acknowledged, that Synercid in clinical trials had not provoked widespread resistance despite the virginiamycin resistance genes that human patients may have absorbed over time. But it *would* happen, Angulo felt, and soon—a lot sooner than would have occurred if virginiamycin had never been used in animals. At that, Carnevale just shrugged and smiled. To him, it was just another case of Chicken Little. Prove it, was his position. And if it did happen, prove beyond the shadow of a scientific doubt that virginiamycin was the cause.

Stephen Sundlof, director of the FDA's Center for Veterinary Medicine, was, like Angulo and Carnevale, a veterinarian. After a childhood of cold Midwestern winters on the outskirts of Chicago, he had fled to Florida, become a veterinary teacher at the state university, and begun to serve on advisory committees for the CVM. His involvement on one committee had led to the chair of another. By 1994, he had moved up to become head of the CVM and pass judgment on new drugs for animal use, as politically difficult and thankless a job as existed in the U.S. government.

Perched in one pan of Sundlof's scales of justice were the pharmaceutical and agricultural industries represented by Carnevale, both bristling at any suggested restrictions on animal-use antibiotics, both wielding lots of political clout. In the other pan sat not just the CDC but a growing number of international groups, including the World Health Organization. Antibiotics given to animals, especially as growth promoters, these groups agreed, made worse a growing global threat of antibiotic resistance in humans. In the spring of 2000, Sundlof heard out the latest round of this debate—the pros and cons of virginiamycin as it might or might not relate to Synercid, now six months out on the market—and did what any bureaucrat in an impossible position would do. He ordered up a study—in government language, a "quantitative risk assessment." Did virginiamycin in fact precipitate resistance to Synercid in humans? Thousands of people would be tested. Information would be gathered. Results would be published. In a year, perhaps.

Or two.

As much as any three people could represent the sprawling complexities of animal-use antibiotics, these three did. But even they were tiny figures lost in a vast, international landscape that could only be defined by numbers. At the start of the twenty-first century, America supported a livestock population of 7.5 billion chickens, 293 million turkeys, 109 million cattle, 92 million pigs, and nearly 1 million sheep, the majority of which were slaughtered each year. Aside from those at a small number of organic farms, nearly all those animals received several antibiotics between birth and slaughter. A pig, for example, received an average of ten different antibiotics. Some of those drugs were given therapeutically to treat individually sick animals. Far more were given therapeutically to whole herds when a few animals were found to be sick: the only way to treat a bacterial infection in chickens, for example, was to put antibiotics into the drinking water of an entire flock. Still more often, whole herds or flocks were given drugs as preventative medicine before being sent to the slaughtering plant so their carcasses would be disease-free. But all these uses paled beside nontherapeutic use. Growth promoters, usually mixed in with food, accounted for as much as 80 percent of the antibiotics given to animals in the United States—24.6 million pounds a year, according to a January 2001 report by the Union of Concerned Scientists. The UCS report also determined that the use of growth promoters had increased by 50 percent since 1985, much of the rise due to a particularly sharp increase in antibiotics by the poultry industry: more than 300 percent in those fifteen years.

Altogether, American farmers used seventeen different antibiotics on their livestock. Of those, thirteen were used for growth promotion. Of those thirteen, seven either were used for human medicine or were in human-use classes: penicillin, tetracycline, erythromycin, lincomycin, tylosin, bacitracin, and, with the arrival of Synercid, virginiamycin. Farmers in the United States used as much and as many of them as they wished without prescriptions. Feed laced with growth promoters was simply bought over the counter. Farmers were also able to buy these drugs as needed, without prescriptions, to treat sick animals. Though Carnevale and his members of the Animal Health Institute were quick to take issue with each next study suggesting the link between growth promoters and human disease, a general truth seemed evident to almost all by the end of the 1990s: growth promoters led certain animal bacteria to grow resistant to those antibiotics, and some of these bugs got passed to people. Two of the three most common food-borne diseases in people, in fact, were getting more resistant to antibiotics each year, thanks in part to growth promoters.

Most notorious of the three most common food-borne diseases was *E. coli* 0157:H7, a newly emergent pathogen that appeared as an asymptomatic colonizer in the intestinal tracts of cattle—a bug that caused neither symptoms nor infections there but caused a range of symptoms in humans and sometimes led to a miserable death. *E. coli*

0157:H7 could be passed not only from animals to humans but by humans to *other* humans—by a salad bar employee, say, who happened to touch lettuce with infected hands. First identified in a 1982 outbreak in a childcare center, it had become a national threat in just a decade, contaminating both food and water. Most commonly, it made its journey to the human gut in undercooked hamburger meat.[2] The bacteria quickly caused vomiting, cramping, and bloody diarrhea. In rare cases, especially with the very young and the very elderly, *E. coli* 0157:H7 could also induce hemolytic uremic syndrome, a life-threatening condition in which the bacteria produced a toxin that caused the red blood cells to break down, and led to acute kidney failure. In September 1999, a three-year-old girl had died from drinking well water contaminated by *E. coli* 0157:H7 at a county fair in upstate New York: the *E. coli* had appeared in cow manure and was transported in rain runoff to a nearby stream, where it infected the well. The following summer, another three-year-old girl died from eating *E. coli*–laced meat at a Sizzler steakhouse in Milwaukee, Wisconsin. North of the border in Walkerton, Ontario, eighteen people died in the late spring of 2000 from town water contaminated by *E. coli* 0157:H7.

All told, the CDC estimated that 73,000 Americans incurred diarrhea each year as a result of *E. coli,* mostly of the 0157:H7 strain, and approximately 1 percent died. But the CDC also determined that most strains of *E. coli* were susceptible to antibiotics: 87 percent of more than 800 isolates of 0157:H7 tested by the CDC between 1996 and 1999 proved susceptible to all antimicrobial drugs. Resistant *E. coli* had been noted, and might well pose a future threat, but so far resistance played an insignificant role in this new pathogen's infection patterns.

---

[2] As an enteric (i.e., gut) pathogen, *E. coli* 0157:H7 was smeared onto cattle carcasses during slaughter as intestines were sawed open along with the rest of the creature. Thus it also appeared on the surface of steaks but was quickly killed in cooking. The grinding of chuck or sirloin for hamburger, however, distributed surface bacteria throughout the meat. A rare- to medium-rare hamburger, as a result, might still harbor live *E. coli* 0157:H7.

Campylobacter and *Salmonella* were both less pathogenic as a rule than *E. coli* 0157:H7, and both were transmitted almost exclusively from animals to humans, not from humans to humans. But both were far more widespread than *E. coli* 0157:H7, and, with both, resistance was very much a concern. *Campylobacter*, like *E. coli*, was a relatively new enteric pathogen, or perhaps an old one identified only in the last two decades. It resided in the intestinal tracts of most warm-blooded animals as an asymptomatic colonizer. In the human gut, however, it accounted for some 2.4 million infections each year in the United States, according to the CDC. Generally, people incurred *Campylobacter* infections from poultry, because a far larger percentage of the bugs (40 to 60 percent) survived the slaughtering process and remained in the end product than in beef and pork (1 to 2 percent). The majority of cases involved moderate diarrhea, abdominal pain, malaise, fever, nausea, and vomiting for one day to a week. Most people who incurred *Campylobacter* did not seek treatment and so did not take antibiotics. But the pathogen could be more serious for small children and the elderly. Occasionally—in about one of 1,000 cases—it caused Guillain-Barré syndrome, a hideous condition of ascending paralysis that could result in respiratory paralysis and death. About 100 Americans died each year from *Campylobacter*, and so the rapid rise of resistance in humans to the drugs given in those cases was disturbing.

Salmonella, also an asymptomatic colonizer in animals, usually passed to humans via undercooked eggs or chicken. In most cases it caused the same symptoms as *Campylobacter*, diarrhea chief among them, and typically laid its victims low for a few days, usually without leading them to seek medical treatment. Rare in the United States a generation ago, it still affected fewer Americans than *Campylobacter*, 1.4 million. But unlike *Campylobacter* it killed about 500. Usually it became lethal when it infected the bloodstream (the condition known as bacteremia). In AIDS patients, *Salmonella* bacteremia was a particularly dangerous threat. And in 2 to 3 percent of patients, the infection led to a reactive arthritis that could be entirely crippling for months, even years. Most of *Salmonella*'s hardest hit victims were among a pool of up to 340,000 Americans infected by a particularly nasty strain resistant to five common antibiotics and susceptible only to one.

Together, *Campylobacter* and *Salmonella* accounted for 80 percent of the food-borne illness and 75 percent of the deaths in the United States. Many if not most of those deaths involved multidrug-resistant strains. But to seasoned observers like Fred Angulo and Glenn Morris and Marcus Zervos, the greatest *resistance* threat from livestock came from another pathogen that caused no obvious human disease at all: animal enterococci. A growing number of scientists felt that enterococci could become resistant to any antibiotic administered to animals for Gram-positive infections, then pass that resistance to human enterococci in the gut. If so, animal enterococci might well transfer widespread resistance in humans to both Synercid and vancomycin, severely curtailing doctors' power to treat the world's most painful and life-threatening bacterial infections.

Stating the case for animal-to-human resistance was easy enough. Proving that a certain sample of meat led to a certain patient's resistant bacterial infection was quite another matter. For decades, Richard Carnevale and his predecessors had had a field day with the variables. An outbreak of multidrug-resistant *Salmonella* in humans might lead investigators to a herd of infected chickens, but who could prove that the resistance had come from those chickens and not, say, from antibiotics the victims may have received over time for other infections? Or, as the lobbyists put it at one point with all due earnestness, from bird droppings that might have fallen on those chickens as they were transported from farm to slaughterhouse?

Certainly the massive use of antibiotics in human medicine contributed much of the selective pressure that led to resistance in people—even the most ardent consumer groups had to admit that. How then was a research scientist to separate the pressure exerted by human-use antibiotics from that which might be exerted by animal-use antibiotics? Even if resistance could be traced to a specific animal source—and it was true that new molecular tools were helping in that regard—how could one determine that that resistance was the factor that led to the disease? Perhaps a victim's immune system was especially vulnerable for other reasons. Perhaps a victim sick enough to be treated at the hospital incurred resistance *there.* Maddeningly, even determining whether patients resistant to a certain drug did less well with that drug was harder than it might seem. Did they take longer to recover from their *Campylobacter* or *Salmonella* infections if their strains of those illnesses were resistant to the drugs used on them? So one would think. And one could assemble, as the CDC often did, epidemiological evidence to suggest this was the case. But saying what happened to 100 patients in a particular hospital was not the same as proving the case scientifically, as Carnevale, the caster of doubt, was quick to observe.

Amid such uncertainty, Carnevale's main points had the luster of unassailable logic. In the fifty years that antibiotics had been used as growth promoters in animals, he declared, there was not one documented case where antibiotic use in animals caused treatment failure in a human patient. How could it, when less than two *ounces* of antibiotics per ton of feed were used subtherapeutically? While some pathogens, resistant or not, did cling to slaughtered and processed meat, Carnevale echoed the simple solution explained right on the package: consumers need only cook their meat thoroughly to kill all harmful bacteria.

These were the arguments that made Angulo grow tense with irritation. Scores of studies had shown that resistant *Campylobacter* and *Salmonella* did transfer to people, and each year hundreds of people got sick or died from those infections, infections that they could only have gotten from food. Insisting on a higher standard of proof than that was like asking for scientific proof that cigarettes caused cancer. Putting two ounces of antibiotics in each ton of feed might seem insignificant, but the fact was that that small amount was enough to engender resistance. Indeed, it presented a perfect opportunity for bacteria: just enough of the drug for the bugs to become exposed to it without being killed and then find a resistance mechanism for it. And in fact, growth promoters often remained in animal guts after the feed that had encased them was digested; then they became much more concentrated and that much more capable of provoking resistance. The overuse of human antibiotics was certainly a cause of resistance, but some of the worst scenarios occurred when the "seeds" of animal-use antibiotics exacerbated that cause: one had only to look at Wolfgang Witte's discovery of human VRE in Europe, provoked by the "seeds" of avoparcin. And if growth promoters were only part of the problem, was that any reason to condone them?

As for Carnevale's suggestion to consumers, that really rankled Angulo. By the CDC's reckoning, 60 percent of the 7.5 billion chickens sold each year in the United States contained *Campylobacter* when they reached consumers. Often the bacteria, originally lodged in the chicken's intestinal tract, were expelled in feces. When chickens were transported to the processing plants in rows of wire mesh cages, the feces from one chicken often fell on the chicken below. Or else during slaughtering the intestines were perforated, and the bacteria were splattered on the carcass. Often the bugs clung to the folds of the dressed chicken's skin or ended up swimming in the juices contained within its retail package. Uninformed consumers had only to handle the raw chicken and then touch and ingest other food without washing their hands, or place the raw chicken on a cutting board and then use the cutting board to chop something else. Then *Campylobacter*—very hardy bugs—would make the jump, followed soon after by a trip down the consumer's esophagus. Declaring that chicken would be contamination-free if it was cooked long enough seemed at best a cavalier stance.

Vigilance in the kitchen, in any event, was no guarantee against bacterial infections spawned by the use of growth promoters on farms. In the meat-packing business it was common knowledge that almost every slaughterhouse worker came down with diarrhea in the first week on the job. Everyone else was susceptible, too, through the environment, where contamination often began with bacteria-ridden feces on farms, especially the megafarms known as CAFOs. Perhaps because the ramifications were so vast and daunting, little had been done to try to measure them. One of the first studies, in fact, had been done as recently as 1999—by a seventeen-year-old high school student.

Ashley Mulroy, of Wheeling, West Virginia, read a shocking story in *Science News* called "Drugged Waters," about antibiotics floating in European waters, and decided that for a school science project she would test her hometown drinking water for three common antibiotics: penicillin, tetracycline, and vancomycin. To her amazement, she found all three drugs in low concentrations. Somehow, no one had thought to do that before. Over a ten-week period, she and her mother also drove to various sites along the nearby Ohio River to collect more samples. Ashley found the same three drugs in all the river samples in significantly higher concentrations. Her findings, which won her an international science prize, suggested that water filtration plants in town failed to remove antibiotics from hospitals and home sewage. (Hospital runoff would account for the vancomycin.) The higher amounts found in river water indicated a free-flow runoff of antibiotics from farms in the area. To that point, a follow-up test by the EPA and the U.S. Geological Survey of surface water and streams near two North Carolina hog farms found antibiotic contamination by chlortetracycline, lincomycin, and sulfamethiazine. The drugs appeared to emanate from wide pools of manure called "lagoons." Swimming in those lagoons were a lot of resistant bacteria. Resistance genes were *everywhere*.

Perhaps, Carnevale conceded, growth promoters on the farm did raise the risk, ever so slightly, that a few people might have diarrhea longer than they otherwise might have from *Campylobacter* and *Salmonella*. But you had to put matters into perspective. Growth promoters helped America's livestock grow faster and kept them healthy. Didn't the nation benefit more from that than by taking an extreme measure to lessen a few cases of indigestion? Wasn't it good that more meat reached Americans; that it was cheaper than it would have been without the large-scale farms that antibiotics made possible; and that human health was boosted by healthier animals? And by the way, Carnevale liked to add, what about the animals? Consumer groups stood ready to attack any possible risk to human health, no matter how minor. But, said Carnevale, leaning back in a black leather swivel chair at the head of a huge polished-wood conference table in the offices of AHI, *who spoke for the animals?*

When he heard that, Angulo's jaw tightened and his eyes narrowed. Congress, he observed, had vested authority in the FDA to allow the use of antibiotics in animals as long as it could provide a guarantee that those antibiotics would have a "reasonable certainty of no harm" for consumers. "It doesn't say 'reasonable certainty of no harm but we'll accept a little more harm if there's more benefit to animals.'"

If these food-borne illnesses were still easily treatable with a wide variety of antibiotics, Angulo added, perhaps resistance in animals would not be a major issue. But they weren't. Instead, the world was seeing the first hints of a future of untreatable infections—coming through the food supply.

In the murky realm of animal-use antibiotics, most of the debate focused on growth promoters. Because they were so widely used and because their possible harm to human health was so considerable yet so difficult to prove beyond the shadow of an industry doubt, growth promoters provoked the most studies and academic papers, the most heated talk at medical conventions, and the most hard-ball lobbying efforts by the agriculture and pharmaceutical industries. But even as Zervos and Morris and Angulo were testing poultry for virginiamycin resistance, a new and potentially *more* alarming threat had arisen from antibiotics proposed for therapeutic use in animals. No one disputed that sick animals needed to be treated. But the drugs that industry wanted to use now were from a relatively new class essential to human medicine, the class that included, among other drugs, ciprofloxacin.

In the nine years since their arrival on the market as human therapy, quinolones and the souped-up fluoroquinolones that followed had become the most widely used antibiotics of their time—and also the first class that could be taken orally, not just intravenously, for certain severe infections. Created in the 1960s by scientists searching for new malaria drugs—chemically inspired, in fact, by quinine, from which the new class had taken its name—they had proved able to inhibit a wide range of Gram-positive and Gram-negative bacteria.[3] Doctors used them to treat urinary tract infections, sexually transmitted bacterial diseases, severe *Campylobacter* and *Salmonella* infections, infections in patients with cystic fibrosis, and many others. The quinolones had proved useful against some strains of methicillin-resistant *S. aureus* (though more useful against methicillin-susceptible *S. aureus*), and in those cases provided an alternative, or backup, to vancomycin. As the world would learn in the aftermath of September 11, 2001, the quinolones, particularly ciprofloxacin, were also drugs of choice against anthrax.

Because the quinolones were entirely made by humans, derived from a synthetic compound called nalidixic acid, they hadn't run up against any long-established natural mechanism of resistance in their targets. Nor could bacteria under attack find resistance genes from any other bacteria. There was, in other words, no horizontal resistance to the drug, only the very rare chance mutation that created resistance—vertical resistance—when bacteria managed to adapt one of their genetic "bilge pumps" to regurgitate the drug.

As Steve Projan, a widely respected expert on the subject, put it, vertical resistance was good in the short term, often bad in the long term. It tended to spread much more slowly than horizontal resistance; that was good. Mutants, moreover, were usually "sick bugs": their gain in resistance meant a loss of something else, so their ecological fitness was compromised, at least until they found a way to compensate for the loss. The bad news was that mutational changes were almost invariably "hard-wired" to a bug's chromosome. If the mutation helped a bug survive, the bug's daughter cells, inheriting that change, would proliferate. Resistance would spread, and might not diminish even if the antibiotic pressure was removed. Worse, despite the rap on mutated bugs as "sick," Projan and others had

---

[3] The quinolones halt bacterial reproduction by inhibiting an enzyme called gyrase, which bacteria need to coil their DNA. As a result, the DNA can't pass along the genetic information for reproduction to the cell's RNA.

found that the mutations necessary for bugs to become resistant to quinolones did little or nothing to compromise their fitness. Those bugs weren't sick at all.

In the long run, then, a wide array of bacteria would very likely develop mutational resistance to the quinolones, and resistance levels in their various populations would rise. But the drugs had a good chance of remaining effective for twenty or thirty years before that happened—*if* they were used sensibly, and not, more to the point, overused.

It was against this backdrop that the FDA's consideration in March 1994 of a quinolone called sarafloxacin, by Abbott Laboratories, for animal use appeared so ill advised. Abbott had applied to have the drug licensed for treatment of chickens afflicted by avian colibacillosis, a respiratory disease caused by *E. coli*. Farmers claimed that no other antibiotic had proved effective against the infection and that hundreds of thousands of chickens died each year as a result. Worse, millions more chickens grew sick and could not be sold. In all, about 2 percent of the roughly 7.5 billion chickens raised in the country each year were being sacrificed to the infection, at a devastating cost. Abbott wanted the drug licensed only for therapeutic use, not as a growth promoter, and seemed agreeable to having it be dispensed by prescription only.

The CDC, at Fred Angulo's urging, sent a letter advising caution in regard to a class of drugs so vital to human health and noted that, despite what farmers said, there was no hard evidence that avian colibacillosis was resistant to all other drugs. In Baltimore, Glenn Morris gave television interviews deploring the proposal. Various consumer groups wrote petitions. At the FDA's hearings on the proposal, held at a dreary Holiday Inn in Gaithersburg, Maryland, academics issued more muted warnings. A few Cassandras declared flatly that quinolone use in poultry *would* engender resistance that *would* transfer to humans. Beginning in the Netherlands, and then in other European countries, they observed, quinolones had been approved for therapeutic animal use over the last several years. Since then, *Salmonella* resistance to the drugs had begun to rise, first in poultry,

then in people. Since people contracted *Salmonella* only from animals, the transfer of resistance appeared clear. In Spain, where veterinarians often dispensed animal-use antibiotics to farmers without
prescriptions, quinolone-resistant *Campylobacter* had been reported.
So had quinolone-resistant *E. coli.* But other speakers waved off the
studies that had come out of Europe and declared transfer not yet
proven. Warily in the middle was Tufts's Stuart Levy, he of the bow
ties and dapper suits whose famous chicken studies had shown that
any antibiotic used in animals would lead to resistance if used long
enough. "You are saying that you need these drugs," Levy said, addressing the industry, "and I am not going to argue that. [But] if that
is true, then it is up to you to see that these drugs are used appropri-
ately. It is not up to us. We can say that if you use them badly, then
you are going to have the trouble and you are not going to be able to
treat those animals that you are telling me are dying."

To these concerns, the industry responded with soothing bro-
mides. In the roughly seven years since the most widely used
quinolone, ciprofloxacin, had come on the U.S. market as a human-
use drug, no full-blown resistance in humans had emerged. Perhaps
resistance could be provoked in animals, as the European stories suggested, but the transfer of resistance from animals to humans was indeed unproven. If quinolones were sanctioned for animal use in the
United States and resistance rates did start to tick upward, in either
animals or people, the industry was willing to be bound by a commitment to yank them from the market. Meanwhile, a lot of sick animals
needed this drug.

Stephen Sundlof heard these arguments with mixed feelings. He
certainly wanted to do no harm to human medicine: that, indeed, was
one of his mandates as the FDA's CVM director. But he had another
mandate as well, to get new drugs to veterinarians. Resistance was a
particularly narrow pinnacle on which to balance those two commit-
ments. Sundlof knew all too well what had come of the FDA's last effort to come to terms with it in the 1970s. Ever since, the agency had
simply ignored resistance as a criterion for judging new drugs.
Instead, it asked only if a drug was toxic to animals and whether it left

any "residue" in humans. Sundlof felt proud that resistance was being brought at last into the debate over a new drug. But that didn't mean he was ready to keep a new, important drug from veterinarians on account of it.

On August 18, 1995, Sundlof declared that sarafloxacin would be approved with chickens and turkeys, though principally for short-term therapy and only by a veterinarian's prescription. To the critics' dismay, farmers would be allowed to dose whole flocks through their drinking water—there was no other way, they had argued, to administer the drug to poultry. But also by Sundlof's ruling, the CDC and the U.S. Department of Agriculture would monitor the drug's effects from the start. Specifically, they would test *Salmonella* and *E. coli* in people and animals to look for evidence of increasing resistance to quinolones. Before, when the FDA approved a drug for animal use, the approval had always been final. Now, Sundlof stressed, it was conditional. If resistance levels rose, the agency could pull the drug.

In his tiny office at the Minnesota Health Department in Minneapolis, an EIS officer named Kirk Smith took in the news of the decision and smiled wryly. He was, he realized, in a unique position to do the best possible test of chickens and quinolones. All he had to do was continue testing isolates his department had begun to record in the early 1990s—isolates that no other state health department had bothered to collect. Now, with quinolones on the market, he'd have comparative figures that no one else would.

Smith knew all too well what that would entail. As Marcus Zervos was testing turkey feces in Michigan, Smith would start testing chicken feces in Minnesota. Glamorous it wasn't: microbiologists would never be movie stars. But Smith didn't mind. He had a pretty good sense that those chicken feces would yield findings as dramatic to his colleagues as any Hollywood thriller to moviegoers. In that, he would be absolutely right.

8

REVOLUTION IN EUROPE

Every weekday morning in Copenhagen, a small yellow postal truck drew up to the Danish Veterinary Laboratory and disgorged a revolting load. Dead piglets, cow heads, sheep lungs, and other offal were carried in sacks to the laboratory's loading dock, then brought inside to be autopsied. Any animal parts found to contain bacterial infections were then tested for drug resistance. Yet until Wolfgang Witte's March 1995 call to Vibeke Rosdahl, deputy director at the affiliated Staten Serum Institute, no one had thought to inspect these fetid organs for vancomycin-resistant enterococci provoked by avoparcin.

When the call came in and Rosdahl directed Henrik Wegener to look for VRE, a lot happened very quickly, for these two agencies, the Danish Veterinary Laboratory and the Staten Serum Institute, formed a government body unlike any other in Europe or, for that matter, the rest of the world. In this complex of eleven buildings, a mix of nineteenth-century wedding cake architecture and boxlike bunkers from the 1970s, the two institutes operated as separate halves of a remarkable whole. Infectious diseases in humans were tracked at the Staten Serum Institute, where Rosdahl, a renowned expert in staphylococci, worked as a deputy director. The Danish Veterinary

Lab applied the same vigilance to infectious diseases in animals. Henrik Wegener, a veterinary professor and researcher, was head of the joint research and surveillance unit that bridged the two: the Danish Zoonosis Center. Because the two were linked, news of a disease outbreak in cattle, for example, would be communicated within hours to the Staten Serum Institute, whose microbiologists might then discern some connection between the animal outbreak and a sudden incidence in human disease. The predilection for animal-use antibiotics to cause resistance was a keen interest in both realms of the institute.

Wegener was a tall, lanky microbiologist in his early thirties with wavy brown hair and green eyes, casually dressed, whose hands moved constantly as he spoke. Buoyant and brimming with good cheer, he nevertheless saw a dark future in growth promoters and had worked for years to rid the world of them by proving that resistance did transfer from animals to humans. Now that he knew to look for VRE as a possible result of avoparcin use, Wegener and a younger scientist, Frank Aarestrup, began by testing meat in slaughterhouses, as well as prepared raw meats in grocery markets, in and around Copenhagen. They found VRE, as Wegener put it, all over the place—in 80 percent of the retail samples they tested. Then they compared conventional poultry raised with avoparcin in their feed to poultry raised organically, without avoparcin. Of eight flocks raised conventionally, five had VRE. Of six flocks raised without growth promoters, none did.

March in Denmark was always the gloomiest time of year, after six months of near darkness but with another month to go before the days began to brighten. For Wegener, though, the light had just come on. "Oh my God," he said to his colleagues, "this is big." Just how big, none of them quite knew as yet. In March 1995, VRE had not yet become a significant problem in European hospitals. To Wegener, the finding of VRE in chickens and humans seemed, more than anything else, an elegant proof of concept that growth promoters fed to animals *could* cause resistance in people. Ironically, the paucity of VRE cases in European hospitals would make the scientists' job of winning a European Union–wide ban on avoparcin that much harder. Why worry when apparently no problem existed?

Denmark had not gone so far as Sweden, which had banned all growth promoters in 1986 even without hard scientific proof. Still, along with the Netherlands, it was just a step or two behind. As soon as the VRE results were in, Aarestrup and Wegener presented them in a meeting to Denmark's leading swine and poultry producers. No hard-ball lobbying ensued, no jockeying in senate subcommittees. Appalled, the producers agreed to stop using avoparcin almost immediately. Within days, Denmark's new and relatively untested minister of agriculture, Henrik Dam Kristensen, declared a national ban on the stuff.

There matters might have stood, except that Denmark was a member of the European Union, and such bans, even if voluntary, had to be justified to a Union committee. The committee would listen to all the evidence Wegener, Witte, and their colleagues had collected and render a decision. Either the ban would be extended to the whole European Union or it would be adopted nowhere, including Denmark. The Danes were told they had six weeks to prepare their case.

Now the hue and cry began in earnest. From Switzerland, the international drug manufacturer Roche, which produced avoparcin, wrote angry letters to the Danish Zoonosis Center, declaring that the ban would create an animal health disaster and that farmers, with sicker and leaner animals on their hands, would go bankrupt. Wegener and his colleagues felt sure that in time these alarms would prove groundless: certainly in Sweden they had. But in the weeks leading up to the EU's vote on avoparcin, fear seemed all too likely to carry the day.

One of Wegener's first moves was to meet with the Swedes. Anders Franklin, head of bacteriology at the National Veterinary Institute near Uppsala, was considered the country's leading authority on bacterial resistance; it was he who had led the move to have Sweden ban growth promoters almost a decade before. With him was a young, very smart protégée, Christina Greko. Eagerly, Wegener explained that what he needed most from the Swedes was data showing a sharp drop in VRE since the 1986 ban, which had included avoparcin. Since the Danes were standing in VRE to up over their heads, as Wegener put it, a striking absence of VRE in Sweden would strengthen the case that avoparcin was the culprit. To Wegener's dismay, the Swedes had no such data. "Have you never gone out and measured the effects of this huge change in animal production practices?" he spluttered. No, the Swedes said with a shrug, they hadn't.

A certain coolness hung in the air between Wegener and Greko, the coolness of two colleagues in a competitive field. Greko, also a veterinarian in her early thirties, was petite and attractive, with red hair and ruddy cheeks, a bit schoolmarmish at first but given, when she felt more at ease, to a rapierlike wit. The daughter of a biology teacher, she had learned the Latin names of plants when she was very young, then become passionate about horses, and combined her interests by going to veterinary school, where she became fascinated by the biology of microbes and the different strategies they chose to survive. After years of teaching at the University of Uppsala, she had just come in January 1995 to work with Anders Franklin at the National Veterinary Institute—a date that coincided with Sweden's entry into the European Union. Sweden had won permission, as a new EU member, to maintain its ban on growth promoters for four years while preparing a scientific case for the EU; Franklin and Greko, a commission of two, would have to make that case.

Wegener's own passion for veterinary science had been shaped by childhood summers on a small, traditional, family farm. But he'd followed a more unconventional path than Greko. In the late 1970s he had dropped out of school to join a commune for what he admitted later were pretty wild years. Though he eventually came back to earn a Ph.D. in veterinary microbiology, he also spent time in southern India, bicycling to the outlying villages of Mysore to teach sustainable development. Toward the end of his stay, he'd met an old Indian professor who had gently mocked his earnest desire to help developing countries. Instead of digging around in the ground with no plan, the professor had suggested, Wegener should go home and do science at the highest level he could. "It doesn't matter what field it is—just do something that is as advanced as it gets in your country. And then, if we need you in the future, we'll call." Wegener had taken the advice.

Now, with no choice but to work together eighteen hours a day, gathering new science and assembling a huge report in just six weeks, Greko and Wegener came to appreciate each other's dedication. They were helped by an extraordinary grant from the Danish government. Hardly any other government had committed funds for the study of antibiotic resistance; because of this grant, much of the science about growth promoters over the next years would come from Denmark. Much of that would be generated by Wegener, helped principally by Greko, Wolfgang Witte, and Anthony van den Bogaard in the Netherlands. Though not organized in any formal way, they would be a gang of four, as Wegener put it, together taking on the entire European Union.

The report on avoparcin was done in time for a European Union committee meeting on July 15 in Brussels, though both Wegener and Greko had their doubts about how it would be received. They'd based the report on the "precautionary principle," a formal EU phrase that gave new evidence the benefit of the doubt. But they knew that the very committee that would judge it was composed of officials who had approved growth promoters in the past. Fueling their anxiety was the knowledge that much more than the Danish ban was at stake. If the committee was unimpressed by their report, avoparcin, along with other growth promoters, would be sanctioned for the foreseeable future in every European Union country except Sweden. Sweden's exemption, or derogation, would expire on December 31, 1998.

Given the gravity of the subject, the EU's Scientific Committee on Animal Nutrition (SCAN) formed a blue-ribbon panel including scientists outside the European arena. The SCAN chairman first contacted Barbara Murray of the University of Texas Medical School in Houston, well known for her work on enterococci. In her letter of reply, Murray reportedly made clear that in her mind there was no doubt that avoparcin produced VRE in animals and no doubt that those resistant bugs could transfer to humans. Disconcerted, the committee instead chose another U.S. expert who espoused the opposite view: all this talk about VRE, he argued, had been exaggerated by scientists and doctors seeking to profit from an "alarmist viewpoint." But when neither this expert nor a British counterpart could produce any credible evidence to shoot down Greko and Wegener's theories, the blue-ribbon panel members found themselves, somewhat to their own surprise, affirming ten declarations, among them that avoparcin selects for VRE in animals; that humans can ingest VRE with their food; and that animal VRE have been detected in the human gut. The declarations were adopted—with a catch. Someone on the committee—Wegener and Greko never found out who—added language to the tenth declaration to the effect that the *magnitude* of the risk remained unclear. Therefore, a formal vote on whether or not to uphold the ban was delayed. Still, Wegener and Greko were highly pleased. Confronted by seemingly closed minds, they had won a critical victory.

With that, the pharmaceutical industry redoubled its lobbying efforts. But for all its money and clout, it found itself fighting a losing battle. In March 1996, the new variant of mad cow disease, called Creutzfeldt-Jakob, erupted, terrifying all of Europe and tilting consensus sharply in favor of the precautionary principle. With mad cow, early warnings had been ignored, with catastrophic results. Why take chances with resistance? From the University of Maastricht, meanwhile, Anthony van den Bogaard conducted a wide study of pigs, chickens, and people in the Netherlands to determine if the ban on avoparcin, also adopted there, had affected their rates of VRE. By the end of December 1996, he would conclude that VRE rates had dropped to half their previous levels.

In person, van den Bogaard was a striking contrast to Wegener and the formal, well-dressed Witte. Bald and heavyset, florid-faced, with his tie askew, and copiously perspiring in the slightest heat, he struck his colleagues as always looking on the verge of a heart attack. Unlike Witte and Wegener, too, van den Bogaard was quick to lose patience with anyone who questioned the link between animal growth promoters and drug resistance in humans. But the rest of the Gang of Four understood his frustration. "What makes you angry is how powerful the argument 'You have no evidence' is in science," Wegener would observe. "Yet our opponents have only questions. They haven't produced any new data that proves us wrong." As van den Bogaard worked closely with Wegener to create a national surveillance network for antibiotic resistance in Denmark—a project funded by the government's munificent grant—Wegener came to admire him as the most unabashedly passionate advocate of the Gang of Four.

When the EU's SCAN met in May 1996 to write up a formal opinion for the EU decision makers, a rancorous debate broke out. Greko's mentor, Anders Franklin, was among those who spoke out forcefully against committee members who seemed inclined to let the "magnitude" issue—Who knew *exactly* how much VRE transferred to humans and *exactly* what effect it had if it did?—sway their opinion. Eventually Franklin and another committee member, Fritz Ungemach, were overruled, but they insisted on having a minority statement be attached to the formal opinion. The opinion was then handed up to the senior EU group that would vote on whether or not the Danish ban should be upheld and extended to all of Europe. The vote came on December 19, 1996.

By the time the EU's standing committee met to vote on whether or not to make permanent the ban, one or another of the Gang of Four had met with every member. Along with van den Bogaard's study, they had Patrice Courvalin's ten-year study of VRE, proving beyond the shadow of a doubt that *Enterococcus faecium* acquired its resistance to vancomycin by importing, and coordinating, highly mobile resistance genes. "These genes were not invented in nature in order to entertain microbiologists," Greko would say. "They are there for microbes to exchange! To say that this does not happen in real life is not biology." Courvalin himself had come out of the lab to do his share of lobbying for the cause, tired, as Greko put it, of all the stupidity among critics who questioned whether resistance genes transferred from animals to humans. What, he would say, do you think those resistance genes are for? To Courvalin, it was absolute nonsense to question whether antibiotic resistance could spread from animal bacteria to human bacteria; it was evident if you looked at the biology. The Gang of Four were hugely gratified to get Courvalin's help, though also a bit daunted. Among microbiologists, as Greko put it, Courvalin was the god. When he walked into a meeting, Greko had noticed, everyone went quiet. Even Witte, the one whose findings had started the campaign, seemed to shrink a little.

Still, from the soundings they'd gotten beforehand, Greko and Wegener felt they could count on only Germany to join Denmark and Sweden in voting to uphold the ban. To their amazement, France swung behind Germany. Spain and Italy, as usual, followed France's lead. With only the United Kingdom and Belgium opposed, the ban was upheld for a provisional period of two years.

Emboldened, the "alarmist scientists" now took on virginiamycin, in view of Synercid's likely arrival on the market. They forwarded their data to Henrik Dam Kristensen, the minister of agriculture, in a report dated January 7, 1998, and waited with bated breath, because everything Dam Kristensen did was a surprise. Of all of Denmark's ministers, he was the only one without an academic degree; he was a mailman from northern Jutland. Initially many Danes had regarded him with little or no respect, but his decision to ban avoparcin had surprised them. Now, after listening to a presentation by Wegener and others in a small meeting room at the ministry, Dam Kristensen paced for a few minutes, muttering to himself, then declared, "I cannot defend virginiamycin to the public." With that, despite enormous risk to his political career, he ordered an immediate ban, which became effective January 16, 1998. As the pharmaceutical industry began lobbying furiously against him, newspaper pundits predicted his imminent demise. Instead, the former mailman became the most popular politician in Denmark. Ordered to Brussels to explain to the EU why Denmark was doing one crazy thing after another with animal-use antibiotics, Dam Kristensen spoke so directly and persuasively that the EU's own officials were won over. In December 1998, the EU adopted Denmark's ban on virginiamycin for all of Europe, and actually extended it, albeit with a two-year review, to all the other growth promoters associated with human medicine: tylosin, bacitracin, zinc, and spiramycin.

In Europe, the struggle over growth promoters was basically over: the "alarmist scientists" had won. And already, fears fanned by the pharmaceutical industry had proven groundless. Without growth promoters, Denmark's poultry suffered no higher incidence of disease. Their weight diminished by a small fraction, which could easily be made up by feeding them a bit more food, the cost of which was exactly equivalent to the cost of feeding them growth promoters. In large swine factories, piglets weaned prematurely from their mothers developed diarrhea without growth promoters, but this was because the food production system had been built on antibiotics; eventually, piglets would be weaned after more careful, appropriate feeding and with better hygiene practices, and their diarrhea would diminish. As in Sweden, no farmers went bankrupt, and retail prices for meat stayed about the same as before the ban. In fact, Denmark became the largest pork exporter in Europe.

In America, the European revolution raised hopes and fears. Consumer groups called for the FDA to follow Europe's lead. Fred Angulo declared publicly that the EU's decisions were based on sound science. Richard Carnevale, his industry nemesis, strongly disagreed. Carnevale had stopped denying that growth promoters provoked some transfer of resistance. The new line was that they did—it just didn't matter. "I'm sure VRE can transfer from animals to people, and it might be resistant," Carnevale said airily. "But is it of clinical importance?"

For his part, the CVM's Stephen Sundlof was under increasing pressure to do something—anything—that might acknowledge growth promoters as a problem, while not alienating the industries that wielded such influence over his budget. Once again, he reached for the fondest solution of all government bureaucrats. In December 1998, just as the EU was announcing its ban on five more growth promoters, Sundlof announced the start of a fact-finding process. He called it the Framework Document. New regulations would be drawn up to classify animal antibiotics in terms of their relevance to human medicine, their likelihood to induce resistance, and so forth. In the highest tier would go drugs both vital to human medicine and vulnerable to cross-resistance. The industry would be hard put to get these drugs approved for therapeutic use in animals, much less as growth promoters. For new drugs in lower tiers, the FDA would reserve the right it had established in licensing quinolones for animal use: the right to yank the drug if resistance levels rose. Details of the Framework Document would be hammered out behind closed doors, but with input from a few public meetings to be held over the next several months.

Or maybe years.

In a sense, as Fred Angulo realized, this was radical action for the FDA. At last, the agency was recognizing that animal-to-human transfer of resistance was a real and serious problem, and that the risk of it should be formally weighed in judging any new drug. So one could make a fairly safe bet that neither Synercid nor linezolid would ever be contemplated as a growth promoter. The Framework Document also underscored that any drug approvals must "do no harm" to human health. That, too, was good. But there appeared to be a gaping hole in the FDA's plans.

The Framework, whenever it was finalized, would apply only to *new* drugs. In a footnote, the FDA advised that it would review already-approved antibiotics only "as resources permit." Its seeming reluctance to broach the subject was understandable. Legally, the burden of proof to prevent the removal of an existing drug from the market lay with the drug maker. By section 512 of the Food, Drug, and Cosmetics Act, a drug maker was obligated to prove that its drug caused a "reasonable certainty of no harm." But this was a gray area. Sundlof had been advised by CVM's chief counsel that in practice the burden of proof actually lay with the agency. If the agency took it upon itself to ban a major pharmaceutical company's profitable drug for animals, Sundlof had better have awfully good evidence, or else lobbying by the angry company would lead to budget-slashing for the CVM by its overseer, the agricultural subcommittee of the appropriations committee of Congress. Just imagining the hearings was enough to make any CVM director break out in a cold sweat. The way Angulo and consumer groups read the Framework Document, no consideration would be given to yanking *any* of the seventeen growth promoters commonly used in U.S. agriculture.[1]

With quinolones, the class of human drug that the FDA had sanctioned for *therapeutic* use in animals, matters had seemed to go from bad to worse. To the CDC's dismay, the FDA had followed up on its 1995 approval of sarafloxacin by approving a second quinolone for therapeutic use in poultry, enrofloxacin, the next year. The agency might well have sanctioned a third, except that by then, as it happened, Kirk Smith had studied an awful lot of chicken feces in Minnesota. What he'd found was shocking.

Typical for an Epidemic Intelligence Service officer, Kirk Smith had started his tour of duty at the Minnesota department of public health in his early thirties. He was a man as unassuming as his name, rather square-faced, of modest height, with short-cropped brown hair and a northwestern drawl that traced to his childhood on a family farm in rural North Dakota. He was, as he put it, "hardly antiagriculture." His paternal grandfather had established the first family farm in the area after emigrating from Poland in the 1930s—hopping trains west through the Depression landscape until he reached the state he liked—and Kirk's father had carried on the tradition, though on a different piece of land, just outside the town of McHenry, population 152. Kirk had grown up doing farm chores, among them dispensing livestock feed laced with growth promoters.

At the state university, Smith studied wildlife biology. But he grew more and more curious about human diseases that came from animals. He went on to earn a master's degree in public health at Iowa State, then attended veterinary school to immerse himself in the

[1]Later, Sundlof would declare, in a very Clintonesque reading of the word *new,* that he'd been misunderstood. "New animal drugs," he would explain, "include drugs from thirty years ago and on up."

other side of the equation. By the time he qualified for the EIS, he felt equally dedicated to protecting animal health and human health. Sometimes, he'd learned, you had to choose one over the other. Then all you could do was get the facts as best you could and let them guide you.

Under Michael Osterholm, the country's best-known and most forward-thinking state epidemiologist, the Minnesota health department in 1991 had established a system for measuring fluoroquinolone resistance in people: a "before" picture that Smith could set against the "after" picture of resistance that may have spiked after fluoroquinolones were sanctioned for animal use. It was a snapshot taken not by intent but by serendipity.

For its system, the health department had looked at human isolates of *Campylobacter* and *Salmonella* to see if either bug was developing resistance to the fluoroquinolones used against it in human clinical medicine. Before 1996, the inspectors hadn't bothered to test for fluoroquinolone resistance in *animal* isolates of either bug, of course, because no fluoroquinolones were yet sanctioned for animal use. Separately, however, they did test animal and human isolates of *Campylobacter* against an antibiotic called nalidixic acid—simply because that happened to be a way to distinguish one species of the bug from another. (The most common species, *Campylobacter jejuni,* usually proved susceptible to it while most other species did not.) Smith saw that those incidental lab tests might be useful in a new light. Nalidixic acid, as it happened, was not only an antibiotic in itself but also a chemical building block of the fluoroquinolones. And a particular property of *Campylobacter* was that when the bug became resistant to nalidixic acid, it became resistant to a whole fluoroquinolone as well. A single "point mutation," as microbiologists put it, created full-blown fluoroquinolone resistance. Thus Smith could look at all those early resistance measurements of animal *Campylobacter* by nalidixic acid as a benchmark for fluoroquinolones as well. By doing that, he might see how much more resistant the animal isolates of *Campylobacter* had become to fluoroquinolones after the drugs were introduced to veterinary medicine in

1994. More important, he might compare this growth of resistance to any similar increase in human resistance and show a link between the two.

A most striking pattern emerged. The rates of nalidixic acid resistance in animal isolates spiked after 1994 as quinolones began to be used in agriculture. So did those of human isolates. But the human story was more complex. Cases of *Campylobacter*—both sensitive and resistant to nalidixic acid—spiked at the end of every calendar year, into January, then steadily declined. Smith's immediate boss, Dr. Craig Hedberg, had an idea about that.

"It's travel," he said. Minnesotans were flying south after Christmas to escape the frigid cold of the Midwest. A lot of them, Hedberg suggested, were going to Mexico. Since 1990, the amount of poultry produced for consumption in Mexico had nearly doubled. The amount of fluoroquinolones used on that poultry had increased fourfold. In a country where public health standards were somewhat lax, chickens were getting inundated with the stuff, which was doubtless provoking resistance in the chickens' *Campylobacter*. The bugs were then transferring to people. For the symptoms of their *Campylobacter* infections—otherwise known as traveler's diarrhea—travelers were in many cases taking fluoroquinolones. In all likelihood, these drugs were also causing resistance to spike in the *Campylobacter* they'd ingested.

Smith had his doubts. "Travel might be part of it," he said. "I don't think it's the whole story, though."

Another contributing factor over the last decade, clearly, was fluoroquinolones used in human medicine for various other infections, both in and out of U.S. hospitals—lots of fluoroquinolones across the board. But while resistance rates in the United States had risen indeed in that time, Smith found, they were modest compared to the January spikes. And they were downright insignificant compared to the period after fluoroquinolones began to be used on poultry.

In all, Smith and his team looked at 6,674 human cases of *Campylobacter* reported to the Minnesota Department of Health, from 1992 through, eventually, 1998. Overall they found a nearly tenfold jump in fluoroquinolone-resistant *Campylobacter jejuni,* from 1.3 to 10.2 percent. For the two-year period of 1996–97—after a fluoroquinolone drug had been sanctioned for animal use in the United States—they interviewed 130 of the patients who had resistant infections, then interviewed a control group of 260 patients with quinolone-sensitive infections.

Most of the patients in both groups, it turned out, did contract their infections from traveling abroad. Whether or not the infection became *resistant* seemed to depend, in part, on whether the patients received fluoroquinolones for their illness. Of the 130 patients with resistant infections, 85 percent were treated with an antibiotic, more often than not a fluoroquinolone. Smith was careful to determine that 20 percent of that 85 percent received a fluoroquinolone before their stools were cultured. That meant that those people might have acquired their fluoroquinolone resistance from the drug itself—exactly what Richard Carnevale would say was the reason nearly *all* resistance to fluoroquinolones occurred. But the rest of those patients had received their fluoroquinolones *after* stool culture, so the vast majority of these patients had to have acquired their fluoroquinolone-resistant *Campylobacter* strains before they were treated with the drug— almost certainly by ingesting those strains in food consumed while abroad.

That was the result that Smith, and more so Hedberg, had expected to see. The revelation was in how many resistant infections began to appear as of 1996 in Minnesotans who stayed put. This group of patients with domestically acquired fluoroquinolone-resistant *Campylobacter* infections—no trips to Mexico, the Caribbean, or anywhere outside the United States—rose from .8 percent in 1996 to 3 percent in 1998, an increase of nearly fourfold. Moreover, these were Minnesotans whose doctors had not prescribed fluoroquinolones for them. No foreign chicken; no drugs at home. All had incurred their quinolone-resistant *Campylobacter* infections in the years directly after the fluoroquinolones sarafloxacin and enrofloxacin were licensed for use in poultry.

Now, Smith decided, it was time to test the other side of the equation. In the fall of 1997, he instructed inspectors from the Minnesota

Department of Agriculture to purchase chicken products from a wide range of markets in the Minneapolis/St. Paul metropolitan area. The inspectors came back with ninety-one products from sixteen retail markets: the pride of fifteen poultry-processing plants in nine states. Some were fresh when bought, others frozen and thawed. Methodically, the inspectors put the chicken parts in big Ziploc bags with a premeasured amount of growth media fluid. They massaged the fluid into the chicken parts to remove the bacteria that clung to the skin. Then they put the fluid in a centrifuge that spun it outward. The residue was scraped off, put on an agar plate, and tested for what it contained. Smith found that 88 percent of the chicken samples contained *Campylobacter*. Twenty percent contained fluoroquinolone-resistant *Campylobacter,* including resistant *C. jejuni* from 14 percent and another strain, *C. coli,* from 5 percent. Nearly all the subtypes of each strain matched fluoroquinolone-resistant isolates from humans.

Here, amid the dry language of epidemiological analysis, was a bombshell. One out of five chickens bought in Minnesota contained a fluoroquinolone-resistant *Campylobacter* infection demonstrably capable of transferring to the people who ate those chickens. And each year, more Minnesotans who ate those chickens were getting quinolone-resistant infections from them.

By now, the CDC had begun its own national surveillance study of fluoroquinolone-resistant *Campylobacter* in people. The last time it had even considered the issue was in 1991. In that year, well after the introduction of quinolones for human therapy, no *Campylobacter* in humans had tested positive for fluoroquinolone—or simple quinolone—resistance. In 1997, not long after the introduction of fluoroquinolones into veterinary medicine, the agency recorded that an astounding 13 percent of all human *Campylobacter* strains tested were fluoroquinolone resistant. The agency also began to test chickens as Smith had done. It found that 60 percent of them, across the United States, contained *Campylobacter*. Of those, about 20 percent were fluoroquinolone resistant.

True, the mortality rate was low—one tenth of 1 percent. Nearly everyone got better. But, as usual with bacterial infections, babies and the elderly were more at risk. And for the rest, *Campylobacter* was no fun at all, even as a self-limiting disease. It was four to seven days of misery and discomfort. Quinolones, at least, tended to end the infection after three or four days, a blessing, especially because some strains of *Campylobacter* had grown resistant to most other drugs. If the fluoroquinolones failed, too, many patients would feel miserable for three additional days. Multiply that by 2 million infections in the United States, Smith observed, and that was a lot of misery, as well as a lot of lost work time and productivity.

To uninformed audiences—such as, perhaps, the Senate subcommittee on agriculture that held the purse strings for the FDA—Richard Carnevale could keep banging his drum that *Campylobacter,* resistant or not, was of little or no account. *Salmonella* was the one that mattered, the serious one, and *Salmonella,* in animals and humans, remained technically susceptible to fluoroquinolones, as it did to virtually every other drug.

But then, to Carnevale's consternation, an odd report wafted over the Atlantic from England. A new kind of *Salmonella* had emerged, out of the microbial blue, one resistant not just to one drug, not to two or even three, but five. This was a strain that caused severe illness and even death, not a few days of diarrhea. One of the only classes of antibiotics that still worked against it was the quinolones.

At first, the trail seemed obvious. On June 18, 1998, Henrik Wegener and his colleagues at the Danish Zoonosis Center learned that cultures from five very sick hospital patients had tested positive for *Salmonella typhimurium* DT104 resistant to five commonly used drugs. Coincidentally, a sample of pork from a slaughterhouse on the offshore island of Zealand had tested positive for the same exact bug. So had two pork samples tested in a routine collection on the mainland by municipal food inspectors. First, Wegener determined that the mainland pork had come from the slaughterhouse on Zealand. Animals from thirty-seven herds had been delivered during the critical three-day period in mid-May. Two of the herds turned up positive for DT104. The herds shared piglets and machinery, which was how the strain had leaped from one to the other. Unfortunately, both the source and the route of transmission remained unknown.

The human cases only deepened the mystery. In the weeks immediately following the initial report, twenty-five cases of DT104 were culture confirmed. All strains matched the resistance profile of the animal samples. But not all victims had eaten infected pork. As Wegener reconstructed it, the epidemic seemed to have begun with a worker at the implicated slaughterhouse. Eighteen of the remaining patients had consumed pork products, including meatballs, tenderloin, and roast pork. But only nine of those eighteen had eaten pork originating from the implicated slaughterhouse. At the same time, a nurse who had treated one of the hospitalized patients contracted the infection. So did a hospital patient situated near the first infected patient. The bug clearly spread by touch or from surfaces, not just by ingestion. It was also very potent. The patients' average age was forty-five—people in the prime of their lives, not the very young and old victims that food-borne infections usually picked on. Moreover, they were sick for an average of fourteen days, far longer than the average food-borne infection usually laid people out, and their symptoms were severe: abdominal pain, diarrhea, chills, fever, and vomiting, as DT104's toxins took over each victim's colon and small intestine then climbed up into the stomach, infecting tissue and obliterating benign bacteria. Most alarming, not all victims responded to fluoroquinolones.

As Wegener noted, the strain of DT104 that afflicted both animals and humans was found to be resistant to nalidixic acid but susceptible to fluoroquinolones, as defined by the bug's MICs.[2] This, however, meant nothing to the four patients on whom fluoroquinolones had no clinical effect. One was a healthy sixty-two-year-old woman whose intestines were perforated by her DT104. Despite treatment with the quinolone ciprofloxacin, along with ceftriaxone and gentamicin, she slipped into multiorgan failure and died. Another of the unlucky ones

---

[2] *Salmonella,* unlike *Campylobacter,* required *two* point mutations to become resistant, the first to nalidixic acid and the second to the rest of the fluoroquinolones.

was an eighty-two-year-old woman with diabetes and uterine cancer whose DT104 also perforated her intestines. Quinolones did nothing to halt her decline and, two months later, her death. The other two patients survived, but their infections were severe and prolonged.

These were DT104's first mortal victims in Denmark, but not in Europe. As early as 1992, a British researcher named John Threlfall in London's Public Health Laboratory Service had noted an uptick in the incidence of multidrug-resistant *Salmonella* DT104. The strain had been around in the United Kingdom for years, he observed, and had grown multidrug resistant in the early 1980s. But not until 1989 had it begun to make a real nuisance of itself. The strain was resistant to ampicillin, chloramphenicol, streptomycin, the sulfonamides, and tetracycline—all drugs commonly used in animals. It seemed to appear first in cattle but soon spread to other species: sheep, pigs, squirrels, raccoons, chipmunks, cats, dogs, and more. In a particularly neat bit of sleuthing, Threlfall found evidence that it might have originated in parrots and parakeets, whose feces then spread the infection to cattle through feed.

Britain's strain of DT104 was, at least, susceptible to the quinolones. But that changed shortly after the quinolone enrofloxacin was licensed for general use in British agriculture. By 1996, 16 percent of the DT104 in animals was quinolone resistant. Given that DT104 had become the most common *Salmonella* strain in livestock, this was a problem. Moreover, the multidrug-resistant strain was killing 40 percent of the cattle it infected. That year, the strain also caused more than 4,000 human infections, a third of them serious enough to require hospitalization, where quinolones proved effective. Most ominously, in a British restaurant outbreak that year, thirteen people were sickened and one died from a strain of multidrug-resistant *Salmonella* DT104 also partially resistant to quinolones.

In the United States, no quinolone-resistant *Salmonella* DT104 had been seen. But the percentage of DT104 in people—strains of *Salmonella* resistant to those five other drugs—had increased from virtually 0 percent among all *Salmonella* isolates in 1993 to 8 percent in 1996. The number of reported cases was estimated to be no more than 5 percent of actual cases, so the CDC guessed that as many as 340,000 cases of five-drug-resistant *Salmonella,* most of it DT104, now occurred each year in the United States.

A first documented outbreak in the United States had arisen among elementary school children in Cass County, Nebraska, a farming community in the east-central part of the state. Between October 12 and 14, 1996, nineteen children came down with diarrhea. In most, fever, headache, nausea, and vomiting also developed. Together, the symptoms suggested *Salmonella.* None of the children required hospitalization, and all soon recovered.

This was the kind of outbreak that state health departments reported routinely to the CDC. Usually these cases were insignificant and merely added to the piles of data that enabled the CDC to arrive at its very roughly estimated annual total of *Salmonella* cases in the United States. This time, the children's cultures tested positive for DT104. The children, it turned out, had been served cold chocolate milk from cartons on October 10. Several cartons of milk with expiration dates prior to October 10 were found in the school's refrigerator, but cultures of samples from the cartons proved negative for *Salmonella* and other enteric pathogens. In England, handling animals had been suggested as another means of infection. Several of the Nebraska schoolchildren had handled a turtle brought in for show and tell, as well as a reportedly ill kitten, in the week before the outbreak. Neither of these animals, however, could be found for testing.

The next U.S. outbreak was both more widespread and more easily traced. In February 1997, researchers at a state microbial diseases lab in Berkeley, California, noticed a spike in *Salmonella typhimurium* DT104 cases from Santa Clara and San Mateo Counties in the San Francisco Bay area—more than fifty of them, all the same strain. All of the victims had Spanish surnames, which soon led epidemiologists to the source: a fresh Mexican-style cheese prepared with raw milk, sold at a local flea market. One vendor and his wife made the cheese in their kitchen from locally purchased raw milk and rennet. The raw milk contained *Salmonella typhimurium* DT104. A second, concurrent outbreak in the area, also from raw-milk cheese, brought the total of victims to 110. By May, a nearly identical outbreak of DT104 in a Hispanic community in Washington State suggested that unpasteurized milk and its use in Mexican-style cheese were fast becoming a major conduit for DT104 in the United States. Raw milk, the investigators observed, was legally sold within certain states, including California, though only to commercial buyers obligated to pasteurize the milk before selling it themselves. As the outbreaks suggested, however, unpasteurized milk sometimes reached consumers.

Farmers drank unpasteurized milk, too—they always had. Milk straight from the cow was one of the perks of the job. But times, and microbes, were changing. Even as epidemiologists in California were testing samples of Mexican cheese that spring of 1997, the members of a farming family in Franklin County, Vermont, were falling deathly ill—one after another.

For Cynthia Hawley, forty-five, a brutal acquaintance with DT104 began when one of her Holstein calves grew feverish and diarrheatic. The calf's eyes were dim, its belly distended. By the time her veterinarian arrived at Heyer Hills Farm, the other calves were sick, too. The veterinarian knew that *Salmonella* was a likely cause of these symptoms. In addition to electrolyte solutions for their dehydration, he gave them ampicillin—usually quite effective against the bug. By the next day, the first calf was dead. So were two others. And the rest had diarrhea that was bloody now, a sign that the bacteria had torn through their intestines. Deeply concerned, the veterinarian sent cultures from the dead calf to Cornell University's veterinary lab. Soon came the preliminary report: *Salmonella typhimurium*. But more tests would be needed before the lab realized that this particular strain was, in fact, multidrug-resistant DT104.

As reporter Amanda Spake recounted in *U.S. News & World Report,* the Hawleys were instructed to stay away from the calves, but no one thought to suggest that they stop drinking fresh milk. The first of the family to fall ill was Nicholas Heyer, five, Cynthia Hawley's nephew. Exhibiting high fever, cramps, diarrhea, and vomiting, he was treated with Bactrim, a combination of two antibiotics, trimethoprim and sulfamethoxazole. Bactrim worked, it turned out, only because DT104 was still susceptible to trimethoprim. Soon half a dozen other family members and farm hands were sick with the same symptoms. All were treated with Bactrim.

Cynthia's own symptoms began some days later, at the hairdresser's. She felt chills, then sharp stomach pains. She asked to lie down, and then found she couldn't get up. Her mother and sister came to get her to drive her car back to the farm. Within twenty-four hours, she had the bloody diarrhea that showed her *Salmonella* was in a more advanced stage than it was in some of the farm's other victims. Rushed to a nearby hospital, Cynthia was given a synthetic penicillin after proving allergic to Bactrim. When that failed, her doctor tried a cephalosporin. That, too, proved useless—a worrisome sign to CDC investigators later, because DT104 was supposed to be susceptible to third-generation cephalosporins. Lying in her hospital bed, Cynthia saw a news report on a virulent new strain of multidrug-resistant *Salmonella* in England. "I know I've got that DT104," she told her doctor. In fact, her doctor had sent Cynthia's cultures to a USDA lab in Iowa and just that morning had received the results. "Yeah," the doctor said a bit sheepishly, "that's what we're dealing with." Now the doctor knew to give Cynthia the one kind of drug she could take to which her strain of DT104 was susceptible: a fluoroquinolone. The one she received was called ofloxacin. Almost certainly, it saved her life.

By the time Cynthia Hawley came home from the hospital after nine days, an EIS officer named Cindy Friedman was at the farm, taking cultures. In all, she found that 22 of Heyer Hills Farm's 147 cows had contracted the infection, and that of those 13 had died. More than a dozen others had tested positive for DT104 but had not become infected. Nearly all the family members who had drunk the farm's infected milk had become sick. But so had a few who *hadn't* drunk the milk— Cynthia Hawley among them. These victims, Friedman concluded, had contracted DT104 simply from handling sick animals. Both routes, she declared, "conclusively showed animal-to-human transmission."

By 1998, Fred Angulo and colleagues at the CDC had determined that a few of the DT104 isolates in the United States were also partially resistant to quinolones and that one was completely quinolone resistant. One case of full-blown resistance out of more than 4,000 isolates might seem insignificant, but as all players in the field knew, rates of resistance to ongoing drug pressure, once established, went only one way—up. In the Philippines, quinolones had been used in agriculture since at least 1987; by 1992, one survey showed 2.5 percent of human isolates resistant to quinolones.

One way or the other, Fred Angulo knew, the quinolone story was about to take a sharp turn. Either the drugs would be pulled from the veterinary market or quinolone-resistant *Salmonella typhimurium* DT104 would spread through the human population, rapidly curtailing the most important antibiotics since methicillin. But even he could not have predicted what that story's outcome would be, and how soon it would come.

On a cold day in December 1999, Glenn Morris made the hour's drive over from Baltimore to the Holiday Inn in Rockland, Maryland, where the FDA always seemed to hold its hearings—the dreariness of the setting perhaps meant to discourage gadflies. Morris had come to testify before Stephen Sundlof and his reluctant commissioners at the second public "Framework" meeting. Along with discussing the Framework in general, the commissioners were due to announce the results of the "risk assessment study" that the industry had demanded to judge whether the veterinary use of quinolones was leading to resistance that transferred to people.

To the surprise of Morris and many in the room, Sundlof began by acknowledging that the study had found that 5,000 Americans in the last year had had a quinolone-resistant strain of food-borne infection. These people had mostly contracted *Campylobacter,* been given quinolones that failed to work, and as a result remained sick two days longer than most of the rest of the 2.4 million Americans who got *Campylobacter* each year. "We have gotten away from talking about zero risk," he said sadly. "We're talking about what is an acceptable level of risk." More than one participant noticed that Sundlof's hands were shaking as he spoke.

That afternoon, Morris was due to sit on a panel that included Fred Angulo. During the luncheon break, Angulo showed him the latest national figures on quinolone resistance, literally just in from the CDC's National Antimicrobial Resistance Monitoring System. NARMS, as it was called, had compiled figures only since 1997. Angulo had just gotten in the preliminary 1999 figure for *Campylobacter* resistance: it had risen from 13 percent of all isolates tested in 1997 to 19 percent. In Minnesota, as Kirk Smith had noted in his landmark study, quinolone resistance in *Campylobacter*—synonymous with nalidixic acid resistance—had measured just 1 percent of all state isolates in 1991.

The implications were obvious.

Morris was a pretty amiable fellow by nature, but those figures made him furious. The afternoon panel had barely begun when he challenged the CVM director. "Did you know this latest NARMS figure?" he asked. "You know what this means, don't you? This is simply intolerable. We have a major problem on our hands, and it's beyond debate. There's only one thing to do about it." Angulo weighed in, too, and as Sundlof twisted in his chair, the pair of them roasted the CVM in no uncertain terms. In the audience, looking no more comfortable than Sundlof, sat the Animal Health Institute's Richard Carnevale. Angulo's subsequent remarks did nothing to ease his stress. Carnevale's latest position had been that *Campylobacter* was a minor infection that didn't matter, but this was quite a rise nonetheless. Carnevale had stressed that no full-blown quinolone resistance had been seen in *Salmonella*, the more serious food infection. But now, in addition to the cases of partial resistance that Wegener had reported, Angulo could report two U.S. outbreaks of full-blown quinolone-resistant *Salmonella*. In Oregon, a patient who appeared to have incurred this infection in the Philippines had transmitted it to eight other patients, first in an Oregon hospital and then in another hospital and two nursing homes. The isolates from all nine patients had the same pattern of multidrug resistance: not only to the quinolones ciprofloxacin and nalidixic acid but to ampicillin, sulfamethoxazole, and trimethoprim-sulfamethoxazole. The outbreak had begun in February 1996 and continued right through December 1999: thirty-five months in which one after another of these nine patients contracted extremely hard-to-treat infections. (In most cases, a third-generation cephalosporin did work eventually, but why, Angulo wondered, did the bug remain on the premises, hovering like a poltergeist? Because, he felt, it was resistant to the quinolones that should have crushed it.) In a Florida hospital, seven patients had come down with a *Salmonella* strain resistant to more than a dozen antibiotics—including quinolones. This outbreak had lasted from July to December 1999. The outbreaks were unrelated, with different strains, and no one had died, but to Angulo and his CDC colleagues they were ominous indeed. Meanwhile, surveys done for the CDC in various U.S. counties revealed that the proportion of *Salmonella* DT104 isolates— susceptible to the quinolones but resistant to five other drugs—had increased from less than 1 percent in 1979 to 34 percent in 1996.

Sundlof offered no indication of what the FDA would do, if anything, and, as winter passed into spring with sphinxlike silence from the CVM, a loose network of consumer group advocates lost patience. Together they demanded, and won, a meeting with FDA Commissioner Jane E. Henney at her office in Rockville, Maryland, to ask why nothing more had been done to yank quinolones from veterinary use. Morris was there. So was Sundlof, who admitted he was in uncharted waters. Legally, there was no established threshold of resistance beyond which he could order the drug companies to pull their drugs. Henney, unfortunately, offered no more encouraging word.

Grimly, in late October 2000, the alliance of consumer groups demanded another meeting with Henney and Sundlof. They wished to report that quinolone resistance was rising yet again and to ask why nothing had been done. This time the advocates filed into Henney's office armed with more facts and statistics, ready to shout and pound tables. This time, Morris noticed, Sundlof's hands were not shaking. As the activists glowered, Sundlof broke into a grin. "Today," he declared, "the FDA is proposing a ban on all fluoroquinolones in animal use."

Morris was stunned, then exultant. With those words, the mood in Henney's office turned from hostility to jubilation. The activists whooped and cheered. Sundlof, the timid CVM waffler, was their hero of the day. And, in truth, this was a heroic decision. It was the first time the CVM had tried to pull any drug on the basis of resistance concerns. In fact, the proposal applied to two: Sara Flox from Abbott Laboratories, and Baytril, a commercial form of enrofloxacin, from the Bayer Corporation.

By the agency's strict protocol, a proposal to ban was exactly that: a proposal. If the relevant drug companies chose to remove their products from the market voluntarily, the proposal would be ratified as written. A company that chose to contest the ban, however, could ask for a hearing. The FDA could say that the facts seemed clear and that no hearing was needed. More likely, to show due diligence and perhaps save itself from a lawsuit, the agency would grant a hearing date in some months', or even a year's, time. The implicated companies would spend that time gathering material to make their case, and the case would then be heard by an administrative law judge who nominally worked for the FDA, though at some remove. Abbott immediately declared it was complying voluntarily. Bayer, with a larger market share to consider, declared it was not.

Given how slowly and deliberately the FDA's Center for Veterinary Medicine would have to move to assure the passage of its most radical action in more than two decades, Baytril could very likely stay on the market, worldwide, for years. It was, after all, a very profitable product.

9

BREAKOUT

A short drive west of Minneapolis on two-lane Route 212, the city's suburbs give way to large dairy farms, their silos, barns, and wood-frame houses set off by miles of running white fences: an undulating patchwork of cultivated squares across the broad Minnesota valley. One of the first true towns out that way is Cologne, which straddles the old Hastings and Dakota Railroad. Cologne has no doctor—it hasn't for decades—in part because Waconia, five miles away, has a good-sized medical center, but also because the townspeople, farming descendants of hardy German, Swedish, and Norwegian immigrants, tend not to get sick. In the winter of 1999, nothing about Cologne suggested it would harbor a new, fast-killing, community strain of methicillin-resistant *Staphylococcus aureus*.

For eleven-year-old Mandy Tice and her mother, Mona, Cologne was a new start. Mona and her boyfriend, Mark, had found a white house with blue trim there that they could afford to rent and moved there with Mandy from a cramped apartment in Waconia. Best of all, Cologne fell within the Waconia school district, where Mandy had been a student since kindergarten. She loved being able to stay in the middle school she knew.

Mandy was a big girl, strong and self-assured, with long, honey-colored hair, a warm and open face, broad shoulders, and a tomboy-ish manner. She loved playing football with the boys, and she had beaten them in a punt, pass, and kick contest that had earned her a trophy. For the most part, she was a healthy girl, too. In the frigid Minnesota winters, as the winds they called Alberta Clippers swept down from Canada, Mandy did seem to get more than her share of common colds and flus, but nothing that soup and bed rest couldn't ease. Certainly nothing had occurred to worry her mother unduly about the symptoms Mandy began complaining about soon after New Year's Day 1999.

Mandy started sniffling and sneezing, so her mother kept her home from school that Friday. Over the weekend, she felt better enough to play outside, but the cold symptoms persisted. On Tuesday night, she had trouble sleeping. Her cough had worsened, and she kept coming downstairs to the bathroom to spit out phlegm. "I'm getting sick of this," she muttered after her fifth or sixth trip down, so her mother gave her a pail to use by her bedside. When Mona woke up in the morning, she poked her head into Mandy's room and saw her daughter wasn't there. Mona thought maybe she was dressed already and downstairs for school, but when she went down, she found Mandy lying in her pajamas on the couch. Mona stuck a thermometer in her daughter's mouth. It came out at 103.9. "That's it," she said. "You're going to the doctor." That was when Mona noticed the pail. It was at least half full. Mona thought at first that Mandy had thrown up flu medicine, because the gunk in it was so reddish in color.

After routine tests and x-rays, Mandy was diagnosed as having pneumonia. Her white blood cell count was high, which suggested she was fighting an infection, so her blood was drawn and sent off to be tested. Then, because her condition seemed to be quickly deteriorating, Mandy's doctor told Mona to drive her immediately to the hospital.

Within a few hours, her blood pressure plummeting and her oxygen level low, Mandy was hooked up to a ventilator in the pediatric intensive care unit of Children's Hospital in Minneapolis. She seemed disoriented and clearly uncomfortable. "I want to go home," she said to her mother. "I'm hungry."

"You have to stay here till you get better, darling," her mother told her, holding Mandy's hand hard.

Mandy looked at her, and, for a moment, her mind seemed to clear. "I'm going to die, Mom."

"Oh, come on, Mandy. You're going to be fine."

"No, Mom," Mandy said. "I'm going to die here. I know it." She gave her mother a look, not of fear or worry, but of knowing what was going to happen.

The ICU doctors wouldn't know exactly what kind of infection Mandy had until the lab results came back, but they took no chances. After heavily sedating Mandy to keep her from pulling out the oxygen tubes from her nose, they gave her a battery of antibiotics, including cefazolin, erythromycin, ciprofloxacin, clindamycin, gentamicin, oxacillin, penicillin, rifampin, and vancomycin.

Dr. Kiran Belani, a specialist in infectious diseases, arrived early in the morning from the University Hospital of Minnesota in response to the news of a strange and serious bacterial case in the Children's pediatric unit. She found Mona and Mark sitting groggily in the pediatric waiting area after a nearly sleepless night on foldout chair beds in an adjacent room. Mona looked at her a moment, searching her memory, then grinned in amazement. "Oh my God," Mona said. "You're the doctor who treated Mandy for that respiratory virus when she was a baby."

Belani remembered Mona, too.

"This is the woman who saved Mandy's life," Mona told her boyfriend. "If there's anyone who can figure out what's going on here, she's the one."

Dr. Belani murmured that she would do her best. In fact, she wasn't hopeful. The culture wasn't back from the laboratory yet, but a quick look under the microscope had confirmed it was some sort of *Staphylococcus*. Given Mandy's severe symptoms, Dr. Belani had her suspicions as to what it might be. "Has Mandy been in close touch with any black or Native American children?"

Mona looked puzzled. "No, I don't think so."

Not long before, a seven-year-old black girl from urban Minnesota had come down with a *S. aureus* infection that was resistant not just to penicillin—by now, that was routine—but to methicillin and all of its successors. To encounter a pediatric case of MRSA was bizarre enough; to find one in which the patient appeared to have spent no recent time in a hospital was unheard of. Everyone knew MRSA was a hospital infection. The girl had entered the hospital already infected, and died after five weeks. Then a sixteen-month-old Native American girl from rural North Dakota had come down with what appeared to be MRSA, too. The sixteen-month-old had died within two hours of arriving at the hospital. Several other cases had involved Native American children, though not with any other fatalities as yet.

Belani had no idea why so many of the earlier cases of community MRSA involved Native American children—whether their ethnicity suggested a link or it was simply coincidence. Nor did she know where the infection had come from or how it had spread. Belani explained that staph infections usually entered the bloodstream through a cut in the skin, but Mona could think of no cuts that Mandy had had. Well, she added, Mona *had* complained of a bad fever blister on her lip. *That could have been it,* Belani thought. Or perhaps the pneumonia itself had enabled *S. aureus* to pass into Mandy's bloodstream: In both Native American cases, a viral influenza had weakened the children's immune systems, especially in their respiratory areas, and that appeared to have given the MRSA bacteria the foothold they needed to pass from the nasal mucosa into the lungs, then into the bloodstream. No one had yet proven that nasal organisms passed into the lungs, but recent findings in the medical journals certainly suggested the link.

Even before Mandy's lab results confirmed that she had MRSA, Belani felt pessimistic. Mandy's lungs had failed, forcing her doctors to put her on a state-of-the-art machine called an extra-corporeal-membrane oxygenator. As the respirator breathed for her, the ECMO, as it was known, removed blood from her body, oxygenated it artificially, and then fed it back into Mandy's bloodstream. Mandy's kidneys had failed, too, so that she was hooked up to a dialysis machine, though even so toxins meant to be excreted by the kidneys were seeping into the rest of her body. Multiorgan failure, Belani knew, was a telltale sign of a violently aggressive *S. aureus* infection, one that was clearly not responding to the antibiotics Mandy had been given.

By Friday morning, when Mona and Mark got the lab results that confirmed Mandy had MRSA, they were no longer alone in the pediatric waiting room. Mona's other children and Mandy's father were there. So were three of her sisters and various other relatives on all sides. Within a day, too, word had gone out to Mandy's classmates and teachers in Waconia, as well as her new neighborhood friends in Cologne, that Mandy was really sick. They had begun showing up in twos and threes, and they simply hadn't gone home. Nearly three dozen people had slept, that second night, in the family waiting room. The hospital wouldn't let anyone except immediate family into the room where Mandy lay strapped in her drug-induced coma, hooked up to all the machines. But they stayed anyway, until the hospital authorities declared the crowd would have to disperse; friends would have to come in shifts. Mandy's friends tried to fight that. "If we can't sleep here," declared one, "we'll sleep in our cars." Finally the authorities herded most of them out and the shifts began, changing every couple of daytime hours: friends, teachers, the school nurse, the principal.

By Monday, nearly all of Mandy's vital organs had failed either partially or completely. Only the machines were keeping her alive. That day, the hospital authorities began letting Mandy's friends go to her bedside in small groups to say goodbye.

Mona didn't know how she'd feel when she told the doctors to turn off the life support machines. In fact, she felt relieved. The doctors had explained in excruciating detail just how hopeless Mandy's condition was. Giving the order felt like the right thing to do, the only choice that she, as a mother, had left to exercise for her daughter.

With the machines off, Mandy died in eight minutes. The date of her death was January 27, 1999.

Dr. Tim Naimi heard about this next case of community MRSA on the day after Mandy Tice was admitted to the hospital. As an EIS officer, Naimi could do no more now for her than any of her doctors. But he could add her case to the big picture he was assembling in his cramped office at the Minnesota Public Health Department and see that it appeared to confirm his worst fears.

At thirty-four, Naimi still looked somewhat like an altar boy, with dark straight hair and clean good looks but modest height. He came across as low-key and genial, almost bashful, but that was deceptive. Like Theresa Smith and all of his other fellow EIS members, he harbored a fierce passion for public health that made it more than just a career choice. As an internist and pediatrician, Boston bred and Harvard educated, Naimi could have graduated by now to a lucrative salary in private practice. Instead, he had worked for the Indian Health Service on the pueblo reservation in Zuni, New Mexico, then gone through the EIS program, beginning his two-year tour of duty in Minnesota in August 1998.

Naimi knew enough about MRSA to be, as he put it later, incredibly startled by what he heard on his arrival about a rash of mysterious MRSA infections among nonhospitalized Native American children around the state. MRSA was a hospital bug. It was like a feasting hyena that saw no need to range beyond the watering hole where its most vulnerable prey gathered. In hospitals, doctors and nurses knew to be on the lookout for it and to knock it out early with vancomycin. But in these nonhospital cases, outpatient doctors had been blindsided by the bug, prescribing cephalosporins or a newer version of methicillin, such as oxacillin, until lab results showed that the *S. aureus* infections they were treating were in fact resistant not just to methicillin but to the entire family of beta-lactam antibiotics, including cephalosporins—the most widely prescribed antibiotics in the United States.

Naimi's next surprise came in studying the blood isolates. Under a light microscope, all MRSA strains looked about the same: like a bunch of golden grapes. But with the test called pulsed field gel electrophoresis, Naimi could determine the genetic makeup of each isolate.[1] Despite the fairly broad geographical region involved—not only northern Minnesota but part of North Dakota—these strains were virtually identical. That was surprising in itself, and scary. Moreover, they were all different from any known strains of hospital MRSA.

Naimi couldn't believe it. A new invader had appeared. Like hospital MRSA, these strains had genes of resistance to the entire beta-lactam family of antibiotics. What they lacked were genes of resistance to the array of *other* antibiotics commonly used in hospitals. The hospital strains of MRSA had acquired genes for these resistances over time from other bacteria—acquired them, as Naimi liked to put it, as so many side dishes. The bug Naimi was looking at didn't have those side dishes, perhaps because, in sensible Darwinian fashion, it had acquired only the resistance genes it needed in the community, where those other drugs weren't commonly used. It was like a filet mignon without the creamed spinach and cottage fries.

How had *S. aureus* done this? How did it spread? And was this a short-lived local aberration or a new epidemic of global implications that had started, for some reason, in Minnesota? Perhaps he was biased, but he had a pretty good hunch about one of those questions. The reason the cases had cropped up in Minnesota, he suspected, was that the state's public health department had such good ties with local hospitals and labs that it could discern patterns of infection more quickly than other states, and because smart medical detectives like Kirk Smith, Naimi's predecessor, and state epidemiologist Mike Osterholm

---

[1]Pulsed field gel electrophoresis, or PFGE, is a vital tool for microbiologists that provides a "fingerprint" of a bacterial strain by analyzing its DNA. The DNA of a strain is sliced into fragments, and the fragments are pushed electrically through a gel, which separates them. As a result of this electrophoresis, as it's called, the fragments can be lined up on a computer screen as bands—sort of like a supermarket bar code.

were there to connect the dots. Even the Mayo Clinic, Minnesota's world-famous hospital, couldn't do that: it lacked the vantage point that public health had.

Of course, that theory might also mean community MRSA had begun to appear in other states and, disguised as random cases of mysteriously resistant flu or pneumonia, had gone unnoticed yet, across the region or even the country.

Medical detective that he was, Naimi went up to the small communities of northern Minnesota where most cases had occurred—many of them in Chippewa Ojibway country—to sift for clues by quizzing doctors and nurses, family members, and, when they were old enough, the patients themselves. He wanted to know, for example, whether a patient's relatives might have had a recent hospital stay, contracted MRSA, and brought it home, even though the strains appeared to be different. He found no such links. He did find frequent cases where one child appeared to have passed the infection to another, or to other family members. Common culprits were suppurating boils or impetigo, common symptoms of *Staphylococcus* that enabled the bug to pass into the recipient's bloodstream. Sheets, towels, or other shared materials could also be the means of conveyance or, as doctors put it, carriage.

Naimi suspected that children were getting the bug more often than parents because they had more physical contact with one another, both in the home and in daycare centers. But this gave him no better idea of where, or how, the bug had originated. He could see, easily enough, that most of the patients appeared to be of humble socioeconomic status and that many of them were indeed Native Americans. But there were exceptions—Mandy Tice, for one, came from a middle-class, not a low-income, family, and even as a general trend the socioeconomic status was a dubious clue. Most often, bacteria became resistant to drugs as a reaction to drugs. Didn't poor people use fewer drugs than wealthy people? Perhaps in some cases, but in others they might use more, if overworked doctors prescribed antibiotics to welfare patients as cure-alls rather than taking time to see if antibiotics were even the right treatment, or if patients passed leftover antibiotics to family and friends to save them a doctor's office fee.

As a standard tactic, Naimi did a case control study, comparing patients who had incurred the new strain of MRSA to patients in the community who had come into the hospital with garden-variety *S. aureus* infections—infections, that is, susceptible to methicillin and other drugs. Had one group taken more antibiotics than the other in the months before infection, or visited doctors' offices more often? Did one have a history of illness the other lacked? Was one group generally older? Made up of more males than females? The answer to all queries was a resounding No Difference. The groups were identical, and so were their symptoms. Those patients with community MRSA just happened to have *S. aureus* that had acquired—from somewhere (Naimi had no idea where)—an extra bit of DNA that conferred resistance to methicillin and other drugs.

As far as Naimi could tell, this was the first time that a community MRSA strain had ever surfaced in America or anywhere else. It might vanish as quickly as it had appeared. But Naimi knew too much about drug-resistant bacteria to hold out much hope of that. In a year or less, this new MRSA could also be spread around the world.

When he got word of Mandy Tice's death, Naimi sent a statewide warning to doctors about community MRSA and then put out the word to neighboring states: Wisconsin, Nebraska, the Dakotas, and Illinois. One by one, these other states began to report cases of their own, all with identical strains. The bad news came from all directions. As his case file grew to more than 300, Naimi also embarked on a statistical analysis. He found that the median age of patients was about sixteen years old, though the ages ranged from one year old to seventy-two years old. The age distribution turned traditional MRSA on its head. The infections were equally distributed by gender, and about 85 percent of them involved skin and soft tissues. Only 5 percent involved pneumonias, but those, as with the case of Mandy Tice, were the most serious ones, along with a smattering of bloodstream and bone infections. Naimi felt as if he were opening the door into a dark room and beginning to see threatening shapes, though the shapes and their relationships to one another remained indistinct.

In February 1999, less than a month after Mandy's death, a fourth child succumbed to community MRSA. The victim this time was a twelve-month-old white boy from rural North Dakota. Like Mandy, he had severe pneumonia, and he had such trouble breathing that he required a chest tube. Also like Mandy, he was initially given routine antibiotics because his doctors assumed the drugs would cure him. On his second day in the hospital, as those antibiotics failed, he had been switched to vancomycin. But for him, as for Mandy, the vancomycin came too late: on the third day he developed severe respiratory distress and hypertension, and he suffered a painful death. When Naimi looked into the case, he learned that the boy's two-year-old sister had been treated for an MRSA infection three weeks before. Isolates from the siblings revealed an identical strain. This was interesting information, though it told Naimi nothing about how the sister had contracted MRSA or why she had survived and her brother had died.

A short walk from Naimi's office on Delaware Street stood the University of Minnesota medical school, and there, in a ninth-floor research lab, worked an obsessive scientist named Patrick Schlievert who felt he did know exactly how the four children of Minnesota and North Dakota had contracted this new strain of MRSA and why they had died. He even felt he knew how they might have been saved.

Schlievert was an expert on *S. aureus*, well enough known that on the day after Mandy Tice reached Children's Hospital he had received a call from one of the attending doctors. As soon as he heard the list of symptoms, Schlievert felt he knew what should be done. But the doctors told him that Mandy was too far gone to respond to the relatively new infection fighter advocated by Schlievert called immunogammaglobulin, a single dose of which cost $2,000. (Dr. Belani reported that immunogammaglobulin *was* used on Mandy but that she was, indeed, too sick to respond to it.)

That same day, Schlievert received a call from a physician in New

York City named Gary Noel. Alarmingly, the call concerned another case of community MRSA, this one afflicting a young boy. Schlievert offered the same advice; the boy was given immunogammaglobulin, and, within days, he walked out of the hospital, his MRSA infection quelled. Schlievert was thrilled by the news—and not displeased to find himself again in the midst of a new, fierce debate about *S. aureus*.

At the university racquetball courts, where he still played a pretty serious game, Schlievert's nickname was "Maverick." A tall, lean, tightly coiled figure in his mid-forties, Schlievert also ran marathons: twenty-four to date, eighteen of which, he was proud to say, he had finished in under four hours. Competitive and aggressive in sports, he had hustled to be in front of the pack in his research, too, and had irked more than a few colleagues as he did. But Schlievert's bold explanation of toxic shock syndrome, widely derided when he'd proposed it two decades before, had proven absolutely correct—air in tampons had allowed *Staph* toxins to proliferate—and so now, as he began to speak out about toxins as the root cause of death in community MRSA, his critics' annoyance was muted, at least, by grudging respect.

As head of his own lab at the University of Minnesota's medical school, Schlievert had followed continuing reports of toxic shock syndrome in various scenarios and concluded that a dozen or more children in the twin cities of Minneapolis and St. Paul were dying of it each year. In nearly all those cases, postinfluenza pneumonia or tracheitis was the official cause of death. But the children also turned out to have the toxin-producing strain of *S. aureus* that had caused toxic shock syndrome. The same specific toxin—TSST-1. as Schlievert had named it—had appeared in 75 percent of the cases, with one of two others, enterotoxin B and C, in the remaining 25 percent. Schlievert theorized that the children's lungs provided the air that those toxins needed in order to grow.

Schlievert studied the blood isolates of thirty-nine patients who had contracted community MRSA, including the four who had died. In thirty-three, he found his old friends, the *S. aureus* toxins. None from this strain, as it happened, were TSST-1. But enterotoxin C appeared in twenty-nine of the thirty-nine cases and enterotoxin B appeared in four of them. To Schlievert, the rest of the puzzle was clear. Three of the four children who had died of community MRSA had suffered from necrotizing pneumonia; the fourth had had respiratory failure. Here in the lungs of all four patients was the perfect breeding ground for *S. aureus* toxins. The rest of the children had survived, Schlievert declared, because they incurred community MRSA as a variety of skin infections.

Vancomycin would have saved the four children, Schlievert agreed, if it had been used in time. But it was too precious a drug to be used as first-choice, empiric therapy, when the toxins had already pervaded the bloodstream, as in Mandy's case, it was useless. So Schlievert had become a huge believer in immunogammaglobulin. Available only since the late 1980s, it was a pool of antibodies from the immune systems of 1,000 human donors. IVIG, as it was called, was generally helpful in any case in which the patient's immune system needed a boost. Specifically for MRSA, the IVIG was sure to contain antibodies to all three of the toxins that produced toxic shock syndrome, simply because its pool was drawn from so many volunteers that one or another would have antibodies from a previous encounter with the toxin in question. The appropriate antibodies from the IVIG would bind themselves to the toxins and inactivate them. That, in turn, would give vancomycin—or a beta-lactam, if the particular strain of *S. aureus* wasn't MRSA—time to help the immune system overcome the infection. IVIG was new, and it was expensive. But Schlievert felt sure it could be the short-term solution to community MRSA.

Tim Naimi, for one, found Schlievert's theory at best hard to prove. Undeniably, the *S. aureus* in these cases had triggered in the patients' immune systems a cascade response that became toxic shock. But whether a toxin or some other intrinsic property of the bacteria had caused the response was, he felt, impossible to say. And almost certainly some host factor was involved, interacting with the bug: after all, not everyone exposed to community MRSA had contracted it, even in cases of an infected child living in the same house with siblings. That the bugs could make toxins was old news, he declared.

What mattered was which toxins were playing what roles, and when? And how did toxins and a whole range of factors conspire to make one person deathly ill but leave the vast majority of others uninfected? "We don't know if toxins are driving the bus," Naimi declared, "or whether the host response is."

Schlievert brushed aside such doubts, as he always had. As a long-time student of *S. aureus,* one of the many in his field who found *S. aureus* the most fascinating and unnerving bug of them all, he marveled at what he saw as its latest means of accomplishing the perfect murder. Other pathogenic bacteria, at least, left glaring clues at the scenes of their crimes. But during humans' brief history on the planet, *S. aureus* had inflicted its varied ills virtually undetected, because it caused so many different kinds of infections. If he were a microbe trying to cause disease, Schlievert mused, he'd do it exactly as *S. aureus* had—subtly, so that it could keep on causing each next disease for a long time before its human host identified the perpetrator. That *S. aureus*'s toxins provoked toxic shock and death did not mean, necessarily, that *S. aureus* intended to kill its human host. It might only want to defend itself against the human immune system long enough to establish itself before moving on to the next human host. But Schlievert, for one, could not imagine a better candidate than *S. aureus,* with its daunting array of infectious consequences, to be the eventual cause of the extermination of human beings from the planet.

10

THE OLD MAN'S FRIEND

Each year, *S. aureus* and *E. faecium* crept into the bloodstreams of tens of thousands of patients around the world, nearly all of them in hospitals. *Streptococcus pneumoniae,* third of the Big Three bacterial pathogens, roamed *outside* the hospital as well as within it. Each year in the United States, according to the CDC, it accounted for up to 135,000 hospitalizations, 60,000 cases of invasive disease, and 8,400 deaths.

In mild cases, *S. pneumo* caused bronchitis, sinusitis, often an ear infection. In severe cases it caused very bad ear infections and a host of invasive diseases starting with pneumonia itself. In fact, *S. pneumo* was the most common cause of community-acquired bacterial pneumonia: community-acquired meaning it originated outside the hospital, pneumonia denoting that ragbag of respiratory symptoms that *could* be either viral or bacterial. It also caused systemic bloodstream infections. In children, it was the biggest single cause of acute otitis media, or earaches—about 6 million a year, according to the CDC—which was, in turn, the most common reason children went to their pediatricians. Now that *Haemophilus influenzae* was contained by a vaccine, *S. pneumo* was also the biggest cause of acute bacterial meningitis, the often deadly swelling of the membrane around the brain, in children. Globally, *S. pneumo* was estimated by National Institute of Allergy and Infectious Diseases director Anthony Fauci to cause 1.2 million deaths a year among infants and young children. And to many experts, growing resistance in a pathogen with such a predilection for children made this the scariest bug of all.

No one could quite understand it, really. After decades of being derided as the "dumb" bug that would never learn how to resist penicillin, *S. pneumo* had wised up in the late 1980s and early 1990s—in biological time, an instant—as if finally provoked by the onslaught of antibiotics hurled against it for so long. In a little more than a decade, it had gone from sleepy susceptibility to global resistance, first to penicillin, then to one drug after another. Even *S. aureus*, historically the bug quickest to acquire resistance to new drugs, looked like a slowpoke in light of *S. pneumo*'s astonishing burst of resistance of the 1990s.

To appreciate the full power of *S. pneumo*, one first had to appreciate the horrifying ease with which it spread: not by contact, as with *S. aureus* and vancomycin-resistant enterococcus, but by airborne droplets. People passed *S. pneumo* by coughing and kissing and by breathing the fetid air of crowded places: nursing homes, homeless shelters, military camps, and, most notably in the 1990s, daycare centers.

As a respiratory tract infection, *S. pneumo* was indistinguishable, until it was examined under a microscope, from viral pneumonia. Indeed, 70 percent of all respiratory infections *were* viral. Attending doctors knew that antibiotics would help only the 30 percent of respiratory infections that weren't. But because they couldn't immediately tell one kind from the other, they did the only sensible thing under the circumstances: they gave antibiotics to *all* their respiratory patients. (In fact, more antibiotics were given for respiratory symptoms than any other kind.) So most antibiotics for respiratory symptoms were misprescribed, which, in turn, accelerated the spread of resistance, as *S. pneumo* in the respiratory tract was exposed to antibiotics, acquired mechanisms to deflect them, and then passed those genes around.

From the respiratory tract, where it most often entered the body, resistant *S. pneumo* was likely to move into the bloodstream unless stopped by drugs to which it remained susceptible. There, the bug could release toxins that attacked vital organs, just as *S. aureus* did, and provoke a life-threatening drop in blood pressure. Severe septicemia, as such systemic infections were called, was lethal chiefly with the very young and very old—as usual, the most susceptible groups—as well as the immunocompromised, such as those with HIV or AIDS. Among hospital patients who had contracted pneumonia of one kind or another in the community—the leading cause of which was *S. pneumo*—one recent study found that the mortality rate was 10 to 25 percent.

In children, *S. pneumo* often brought on a bad earache instead of a respiratory infection. Then it confronted the pediatrician with a similar dilemma. The truth was that most ear infections in children older than 24 months were clinically mild and healed just fine *without* antibiotics. But how many doctors could refuse a prescription to a child in pain, especially one whose anxious, demanding parent was standing by? Under the onslaught of antibiotics, many strains of *S. pneumo* in otitis media grew resistant. And so the earaches went away for a while and came back, more resistant each time. Up to 61 percent of the children in some daycare centers had been found to be colonized or infected with drug-resistant strains of *S. pneumo* in otitis media.

No amount of vigilance, it seemed, could necessarily protect a child exposed to other children, all with still-vulnerable immune systems, in such close quarters, day after day. Susan Donelan, infectious diseases attending physician and director of infection control at Stony Brook Hospital in central Long Island, New York, brought all her professional standards to bear in choosing whether to send her ten-month-old son, J.P., to a local daycare center in the spring of 1998. Was the bathroom kept clean? *How* clean? Were the refrigerated lunch and snack foods fresh? Was the date on the milk in the front of the refrigerator closer to expiration than the date on milk in back? Were the people who prepared the food different from those who served it? And did those who prepared it wear disposable plastic gloves? The daycare center passed Donelan's test with flying colors, and she brought J.P. there for the first time on May 18. Two weeks later, his earaches and high fever began. Each time, Donelan gave him children's Motrin and Tylenol, along with a ten-day course of antibiotics prescribed by his pediatrician. The symptoms would subside, then reappear. On the night of July 1, J.P. awoke with a temperature of 105 and a swollen ear that looked all too conspicuous by morning.

Donelan and her husband, John Dervan, a cardiologist at Stony Brook, were lucky in a sense: J.P. suffered no permanent damage to his hearing. But at his parents' hospital, the boy's swelling was diagnosed as mastoiditis—inflammation of the temporal bone behind the ear—which was serious indeed. In surgery the next day, a half-moon incision was made behind J.P.'s ear in order to debride out pus and part of the mastoid bone. Four days later, he went home on a course of ceftriaxone, the antibiotic that had proven most effective against his particular strain of multidrug-resistant *S. pneumo*. The drug had to be administered by injection, intramuscularly. Susan could have done it herself but insisted that a nurse do it instead. "Mommies aren't supposed to hurt," she put it.

From his ongoing exposure to other children at the daycare center, and from being supine much of the time so that bacteria spread all too easily from his throat, J.P. had recurrent ear infections. Finally, his parents agreed to have small draining tubes surgically inserted in J.P.'s ears. This was the procedure, all too common with recurrent pediatric ear infections, of actually perforating the eardrums and implanting in each a small tube to drain the frequent buildup of fluid from infection. The very thought of it was enough to make any parent shudder.

After about three months, the tubes were removed and J.P.'s eardrums appeared to heal. But the legacy of J.P.'s resistant *S. pneumo* continued: more earaches, more infection. Not for another year would the pattern appear to subside—a consequence, his mother felt, of J.P.'s maturing immune system and the fact that he now spent the whole day standing up. Even so, the odds were that he would be more vulnerable to infections in the future than would other children and that those infections would be harder to dispel with antibiotics.

Dumb as it had seemed in its susceptibility to drugs through most of the last half of the twentieth century, S. pneumo had always been a worthy microbial foe. Unlike other kinds of bacteria, its cells are enclosed within a hard capsule made of complex sugars called polysaccharides. Subtle differences in those sugars produce ninety varieties, or serotypes, of S. pneumo bugs, each just different enough from the others for the immune system to have to learn to identify it in order to target it. The capsule helps make it virulent, enabling S. pneumo to adhere to the lungs' mucosal linings, which is how the bug starts to cause pneumonia by producing lung-clogging mucus.

Before the age of antibiotics, S. pneumo was "the old man's friend." It and other causes of pneumonia spirited away an elderly, infirm patient with a minimum of pain and suffering—a natural form of euthanasia. Soldiers died of it in wartime, either as a complication of battlefield wounds or as the result of prolonged exposure to the elements. Small children, their immune systems as undeveloped as their grandparents' were compromised, died in droves from pneumonia, too. All told, in the opening decades of the twentieth century, S. pneumo and its fellow causes of pneumonia took more lives than any other disease except tuberculosis. Then, in the midst of World War II, came penicillin. Almost overnight, mortality rates from S. pneumo dropped from 30 percent to 5 percent and stayed down, lulling doctors and drug companies into complacency. Which was where they remained until a frightening outbreak in South Africa.

One day in the spring of 1977, Peter Applebaum, a South African–born physician serving at that time as head of the microbiology lab at Durban's King Edward the Eighth Hospital, analyzed a routine blood culture from a four-month-old black infant who had come in with fulminating meningitis and a perilously low white blood cell count. The attending doctors treated her with everything they had at hand: penicillin, kanamycin, chloramphenicol, and more. Six days later, the girl was dead, an all-too-typical outcome of infection in a woefully malnourished infant. But in Applebaum's lab her *S. pneumo* was still growing on an agar plate. On a whim, Applebaum ordered the resident doing the cultures to test the strain against penicillin. "Why?" the resident asked. "We know *S. pneumo* is susceptible to it." Applebaum told him to test it anyway.

To the resident's surprise, and to Applebaum's, the strain grew, uninhibited by penicillin.

Applebaum started poking around in the medical literature and learned of a few scattered reports of penicillin-resistant *S. pneumo* over the previous decade. The first had come from New Guinea in 1969, cited by a Dr. David Hansman. But these were in carriers, not clinicals, as the jargon had it: people who were colonized but not outrightly infected. None had actually gotten sick and required antibiotics. A few more anecdotal reports had come from the United States and England, as well as from a small town in southern Romania, all in the early 1970s. These were isolated cases, however. Freak occurrences. Meanwhile, in the lab, Applebaum began testing all new *S. pneumo* isolates against penicillin. He found a second case, then three more.

What Applebaum was seeing was the first clinical outbreak of penicillin-resistant *S. pneumo,* the beginning of *S. pneumo*'s reawakening from "dumb bug" to dangerous killer: a strain resistant, as it turned out, not merely to penicillin but chloramphenicol, the two drugs used to treat bacterial meningitis in Africa.

Applebaum sent his isolates over to Johannesburg to be verified by the head of the microbiology lab at Baragwanath Hospital in Soweto. Baragwanath was a huge, 2,000-bed institution, probably the largest in Africa, affiliated with the South African Institute for Medical Research, the biggest lab in the country. Sending isolates there for verification was good medical protocol. It would also serve as a warning to the hospital. Perhaps if the institute's Dr. Henrik Koornhof started looking for penicillin-resistant *S. pneumo,* he would find it, too. That would save lives, even as it bore out Applebaum's suspicion that an outbreak of truly alarming implications was occurring right under their noses.

Koornhof confirmed the finding and shared the startling news with a young doctor named Michael Jacobs, who had just arrived at Baragwanath to train with Koornhof and take over the hospital's microbiology lab. Jacobs was amazed. This was, as he put it later, nothing less than earth-shattering. He started testing isolates from the pediatric ward, where more than sixty fever-stricken children with not very good hygiene were being kept two or more to a bed, all of them coughing and sneezing on one another. Three weeks later, he found a first isolate resistant not only to penicillin and chloramphenicol but also to the macrolides, the tetracyclines, and trimethoprim-sulfamethoxazole.

Whether or not this was the first case at the hospital—the index case, as doctors put it—remained unclear. Perhaps, as in Durban, it had been infecting children for some time, unnoticed. All Jacobs could say for sure was that the bug spread with terrifying speed. Within weeks, nearly every one of those children had incurred the same multidrug-resistant strain.

As Jacobs studied the cases, he did find a common denominator. Most of the children had contracted measles first. In one typical case, a child who had come in for elective cardiac surgery caught measles in the hospital, was sent off to a fever hospital to recuperate in isolation, then returned for his surgery. Postoperatively, he developed multidrug-resistant *S. pneumo*. The strains seemed confined to the hospital at that point, which was odd, since *S. pneumo* traditionally resided in the community and passed easily through cough- or sneeze-expelled droplets. But perhaps, Jacobs came to feel, young children in the community had been exposed early on to mild versions of these strains and had grown immune to them. Very young children in hospital, by contrast, were immunocompromised. If they then contracted a viral infection like measles, it would damage the respiratory tract and provide a fertile field for *S. pneumo* to take hold. The worst cases involved meningitis, for which penicillin plus chloramphenicol had long been an inexpensive therapy of choice. Actually, combining chloramphenicol with ampicillin worked well, too. The problem with these *S. pneumo* strains was that they were resistant to penicillin *and* chloramphenicol. So then neither combination worked. The best fallback was vancomycin, but that drug was far beyond the means of a poor country like South Africa.

As the outbreak spread, the researchers contacted David Hansman in Australia to ask for a sample of his decade-old strain of penicillin-resistant *S. pneumo*. The strains were different—Hansman's had been resistant only to penicillin—but close enough to view the New Guinea case as a precursor of the South Africa cases. Before long, a similar clone turned up in Spain, then radiated out in all directions. As it did, it also broke out of the hospitals, colonizing and ultimately infecting children in the community. For some unknown reason, the South Africa/Australia clone failed to spread from those countries to others. But nor did it surrender any of the ground it had gained. Twenty-five years later, it remained an ineradicable resistant bug. Together, the two clones had colonized the world.

How, Applebaum and Jacobs marveled, did the bug *do* this? What mechanism of resistance had it found that counteracted penicillin and other drugs so well, one after another? And how did resistance spread so easily from country to country? To answer that, Koornhof and Jacobs needed a master geneticist. To Applebaum's enduring irritation, they called Alexander Tomasz.

Tomasz had just become a professor at New York City's Rockefeller University, a brilliant young heir to the lab of Oswald T. Avery, Colin M. MacLeod, and Maclyn N. McCarty. He worked in the same high-ceilinged, sunlit rooms of an ivy-covered building at the north end of the campus where those three had used *S. pneumo* to prove that DNA was the stuff of heredity. Tomasz was a sharp contrast to that Anglo-Saxon trio. He had come as a young man to America from Hungary in 1956 after fleeing the Soviets in the back of a farmer's truck, with no money or possessions. Eager to become a microbiologist, he had written a letter, as a refugee in Austria, to Joshua Lederberg of Rockefeller University. Lederberg had helped find him a job as a technician at Sloan-Kettering Hospital; from there he had obtained a Ph.D. at Columbia and had come to Rockefeller to work for one of Avery's research partners, Roland Hotchkiss. With Hotchkiss, he had delved into the study of S. pneumo's genes and polysaccharide capsules, the better to understand how it became resistant to antibiotics.

By 1979, when Koornhof and Jacobs sent him a first batch of isolates from the South African S. pneumo cases, Tomasz was a well-known Cassandra on the burgeoning threat of antibiotic resistance, specifically with S. pneumo. Colorful, dramatic, moody, and brash, he was an outsize character in the often muted company of fellow microbiologists, warning his colleagues, in his thick Hungarian accent, that the end of the antibiotic era lay just ahead. Applebaum, for one, would come to feel that Tomasz was a grandstander who capitalized on the South Africa findings and took too much credit for himself. But then, more than one of Applebaum's colleagues would say he was a hard man to please.

When Koornhof described the isolates he was sending, Tomasz assumed he would see that S. pneumo had managed at last to do what S. aureus had done decades before against penicillin: imported a gene that encoded for an enzyme that could sunder penicillin's chemical rings. Perhaps the South African strain of S. pneumo had even acquired a penicillinase, as these enzymes were called, directly from S. aureus.

To Tomasz's surprise, S. pneumo had adapted a far more elaborate method, one he had never seen before. Using DNA imported from some other bug, the penicillin-resistant strain from South Africa had actually redesigned the cavities of the cell into which the drug fitted and "retrained" its "workers" to build the cell wall using that redesigned cavity. Tomasz liked to compare the challenge to that of a rogue state like Iraq buying complex new weapon systems from Russia. Just buying the technology would accomplish nothing. Iraq would need Russian technicians who could set up the systems and maintain them. With its new cavities and worker instructions, the South African strain of *S. pneumo* had made a huge, terrifying leap.[1]

At the same time Tomasz was studying the South African clone, a very smart colleague in Spain named Herminia de Lencastre—whom Tomasz would later marry—began finding ample evidence of another, far more widespread clone. In Spain, where over-the-counter sales of antibiotics were all too common, resistance was worse than in any other corner of Europe. Soon de Lencastre realized there were two closely related clones, a Spanish-USA clone, as she put it, because of its geographical spread, and a French-Spanish clone.

De Lencastre was among the first to observe that children in day-care centers were especially vulnerable to these clones of multidrug-resistant *S. pneumo.* Before daycare centers, small children had been kept at home, exposed to other small children one or two at a time. Now they were thrown in with twenty or more in one confined space, day after day after day: touching one another, wrestling with one another, sharing spit and mucus, coughing and sneezing on one another. De Lencastre found that 50 to 60 percent of the children she studied were colonized with *S. pneumo,* vastly increasing the odds that a child with a cold, his or her immature immune system repressed, would acquire a full-blown *S. pneumo* infection that might, under a barrage of antibiotics, become multidrug-resistant. From the perspective of public health, de Lencastre began to write, it was an absolute, unmitigated disaster.

And there was no sanctuary, it seemed, from this new bug. Iceland, where strict government health protocols and a homogenous population on a remote island kept *S. pneumo* cases to a minimum, with resistance rates at 0 percent, saw its first case of penicillin-resistant *S. pneumo* crop up in 1988. Despite government vigilance, 17 percent

---

[1]One theory was that *S. pneumo* had acquired the DNA from *Streptococcus viridans,* a human commensal that lives year-round in the throat. Pneumococci do not have plasmids, which would have made gene swapping particularly easy. But *S. viridans* could have passed its DNA for multidrug resistance on a transposon, or by transformation.

of all *S. pneumo* on the island was penicillin resistant by the end of 1992—an astounding leap. Karl G. Kristinsson, a microbiologist at Reykjavik's National University Hospital, sent isolates to Tomasz in New York and to another distinguished figure in the study of *S. pneumo,* Dr. Robert Austrian of the University of Pennsylvania. Tests by all three experts confirmed that the majority of those isolates had the same pattern of *multi*drug resistance, resistance not just to penicillin but to tetracycline, chloramphenicol, erythromycin, and trimethroprim-sulfamethoxazole. After much medical detective work, Kristinsson determined that an Icelandic travel agent had vacationed in Spain and come into contact with a Spanish child colonized by the Spanish clone. The agent had been exposed to droplets from the child and become an unwitting carrier. Back home, he had coughed and sneezed on his children. They had become colonized, then spread the clone through their daycare center.

Already, the Icelandic health service had prudent rules in place for antibiotics. Now the rules were made even more stringent. Use of all the drugs to which the Spanish clone was resistant was sharply curtailed. Without the selective pressure from penicillin and those other drugs, Spanish clone bugs encumbered by their extra chunk of DNA for multidrug resistance no longer had the ecological advantage over their nimbler, susceptible brethren, and so in the world of *S. pneumo* a radical demographic shift took place: susceptible bugs prevailed. Four years later, in Iceland's tightly controlled environment, the rate of multidrug-resistant *S. pneumo* was back down to nearly zero.

Like Iceland, the United States initially seemed exempt from multidrug-resistant *S. pneumo*. A national survey by the CDC from 1979 to 1986 detected intermediate penicillin resistance in 5.1 percent of about 5,000 isolates but full-blown resistance in just one strain, or .02 percent of the whole group. Then, in 1993, reports began to appear from Kentucky, Tennessee, and other states of an alarming spike in penicillin resistance among schoolchildren. The CDC would determine that by the late 1990s, nearly a third of all *S. pneumo* isolates in the United States were penicillin resistant. In the Baltimore area,

Glenn Morris found that the highest rates of penicillin resistance appeared in the wealthy suburbs. Why? Because, he theorized, those children were constantly being taken to their pediatricians and given antibiotics, penicillin chief among them, so that resistant strains predominated that much more quickly.

Actually, by the late 1990s, the term penicillin-resistant *S. pneumo* was nearly quaint. As often as not, a strain proved resistant to a host of other drugs, too. One of the few classes that remained effective against it were the quinolones. But Don Low, who had fought so hard to contain an outbreak of VRE at Ontario's Mt. Sinai Hospital, chilled the field with his 1999 report that fluoroquinolone resistance in *S. pneumo* in Canada appeared to have risen from 0 percent in 1993 to 1.7 percent in 1998, a huge percentage leap that augured worse to come. Penicillin resistance in *S. pneumo* had taken twenty-five years to emerge as a clinical problem, even in rare instances, after the introduction of penicillin. Quinolone-resistant *S. pneumo* had become a problem in Canada just a decade after the drugs' introduction.[2]

For the few but ominous cases of quinolone-resistant *S. pneumo* beginning to appear, a doctor could always turn to the last-resort drugs used on *S. pneumo*'s fellow Gram-positive pathogens. So far no resistance to Synercid and linezolid had been noted in *S. pneumo*. Of course, the drugs had been out on the market only for a year or so. With vancomycin, the picture was more complicated. Clinical isolates of *S. pneumo* had been identified that were *tolerant* to vancomycin — a phenomenon first noted by Tomasz, in which bacteria seemed to play possum in the presence of a drug, choosing not to grow and thus not exposing their cell-wall-building targets. Those strains were also penicillin *resistant*. Bill Jarvis of the CDC assumed the next step was inevitable. "The concern is that [penicillin-resistant] *S. pneumo* will eventually become vanco resistant, and when that occurs, it is going to be a big problem."

No one could say why *S. pneumo* hadn't yet acquired full-blown

---

[2]Both quinolones and another class of drugs, macrolides, became resistant to penicillin using efflux pumps rather than changing their cell wall proteins.

vancomycin resistance. There was no good reason, for example, why VRE could not pass on its vancomycin-resistance genes to *S. pneumo* across species boundaries. Perhaps, as Tomasz mused, billions of *S. pneumo* bacteria did acquire vancomycin resistance genes every day but incurred some cost in fitness that kept them from surviving more than a nanosecond. In any event, vancomycin was at best an awkward fallback for community *S. pneumo,* as it was administered intravenously in hospitals at considerable cost.

In the long term, *S. pneumo* experts like Alexander Tomasz and Don Low put their highest hopes in vaccines. Because the *S. pneumo* bacterium was encased in a polysaccharide capsule, a vaccine could use elements of the capsule, detached from the bacterium's DNA, as an antigen: these capsule bits would provoke the human immune system to make antibodies against *S. pneumo* without posing the threat of the actual bacteria. If the vaccinated host was hit with a real *S. pneumo* infection later on, his or her immune system would have antibodies tailor-made to target it and would not waste life-threatening days manufacturing them for the first time.

A first vaccine marketed in 1977 contained as antigens fourteen different polysaccharides—different complex sugars of the pneumococcal capsule—which blocked many, but not most, of *S. pneumo*'s virulent serotypes. An improved version contained twenty-three polysaccharides and covered nearly all the virulent serotypes. Unfortunately, children under two years old could not be vaccinated: their immune systems were too immature to manufacture antibodies in response to the vaccine's antigens. In early 2000, Wyeth-Ayerst came out with a "conjugate vaccine," marketed under the name Prevnar. The conjugate, or combination, of the two antigens elicited an immune response of T cells and production of certain antibodies even in small children. Most dramatically, the new vaccine all but eliminated pneumococcal meningitis. As a result, declared its manufacturer, the lives of four infants and toddlers were saved in the United States each day. Of course, as with any vaccine, Prevnar had to be given to a child as a preventative measure, before the immune system came into contact with the bug.

For those children fortunate enough to be vaccinated early on, multidrug-resistant *S. pneumo* might pose no more of a threat than chickenpox or measles to a child inoculated for those once ravaging infections. But for all too many children, a pathogen impervious to all the usual antibiotics now required treatment with one of the quinolones or even vancomycin. The costs involved were startlingly higher than for treatment with penicillin, and the quinolones were still considered dangerous for children. In any event, greater use of these last-resort drugs was sure to increase rates of resistance to them in the fast-evolving strains of the bug no longer viewed by anyone, anywhere, as dumb.

S. pneumo had become a global bacterial threat at the start of the twenty-first century—a threat to all countries but especially alarming in developing nations, where the costs of those few drugs still effective against multidrug-resistant strains was prohibitively high. Rosamund Williams of the World Health Organization, assigned the daunting task of establishing a global resistance monitoring network on a shoestring budget of less than $2 million, flatly called *S. pneumo* the bug she feared the most.

At least *S. pneumo*'s symptoms still gave doctors enough of a warning to treat the majority of cases successfully—*if* they had the drugs they needed. To an expert like Ontario's Don Low, another kind of streptococcal infection presented, on a case-by-case basis, a far more alarming scenario: symptoms so subtle as to be almost unrecognizable and a brutally quick, incredibly painful, almost always lethal course of infection.

11

FLESHEATERS

For Evangeline Ames Murray, forty-four, a recent mother of twins who lived in the affluent bedroom community of Darien, Connecticut, a sudden, unexpected acquaintance with necrotizing fasciitis, better known as flesh-eating bacteria, began as a stinging sensation, like that from a paper cut, on her left index finger at about 9:00 A.M. on November 29, 2000. Evangeline was on a commuter train to New York, a later one than she usually took to get to her office in Manhattan, because, ironically, she was accompanying her husband to a hospital for sinus surgery. Disembarking, she noticed a small red puncture mark on her finger. Later, her doctors would theorize that Evangeline had been bitten on the train by a brown recluse spider, a usual resident of warmer climes that often found its way north amid bags of mulch chips, then in the fall sought cozy nesting places like cellars—and trains. But the spider was merely a chance player in the drama about to unfold. Its bite had provided an opening in the skin at a site occupied at that particular time by some very, very dangerous bugs.

During the five hours that her husband was undergoing treatment at the hospital, Evangeline sat in the waiting room feeling the onset of what seemed a bad flu. At the same time, her little "paper cut" grew much more painful. Evangeline prided herself on having a high threshold of pain—not only had she borne twins, but she was a cancer survivor. This was *intense.* Finally, she flagged down a doctor hurrying by and asked him to take a look. "That's a cellulitis," he said, using the medical term that meant an inflammation of cellular tissue, "and it looks like it's ascending."

Evangeline decided to tough it out, though when her husband emerged from his five sinus procedures, she said she couldn't face the commuter train again. Her finger was hurting too much and turning a strange color. Instead, she called a car service. Her husband, in considerable pain himself, was more than willing to splurge on a chauffeured ride home. A grim day turned worse, however, when their nanny opened the door to inform them that Evangeline's father-in-law had died that day of a heart attack.

Despite the news and her husband's condition, something told Evangeline to go to the emergency room of Stamford Hospital about her finger that night. It was an instinct that saved her life. She arrived at eleven P.M., and was looked at by a nurse who had pretty good instincts himself. Though he'd never seen a clinical case of necrotizing fasciitis, the nurse felt that Evangeline's finger looked worrisome enough—by now it was turning black—to merit a late-evening call to the hospital's head of infectious diseases, Dr. Gavin McLeod. As soon as he heard the symptoms, McLeod called Dr. Chan Rha, a hand surgeon. The two hurried in to find the blackness in Evangeline's finger spreading to her other fingers. The pain, Evangeline told them, was excruciating. That night, they performed a first debridement, cutting away dead flesh from Evangeline's finger in the hope that new, healthy flesh would grow in its place.

For the next ten days, Evangeline's life hung in the balance. She lay immobilized in a hospital bed amid intravenous lines for morphine and antibiotics, her blackened hand suspended from the ceiling in an effort, the doctors explained, to help keep the "bad blood" sequestered from the "good blood." So virulent was the bacteria, however, that regular doses of several drugs, including vancomycin, seemed to have no effect. Nor did any of the three debridements that

Rha performed in all. To cheer her up, the nurses who changed her bandages would tell Evangeline she was one of the lucky ones. "We saw one other person with this infection in Stamford Hospital," they told her, "and he went home with no arms and no legs."

After ten days, Rha announced, "I think we've stopped the eating."

As a precaution, Evangeline was kept in the hospital twelve more days. As the pain subsided, she grew increasingly restless. "In the time I've been here," she complained to McLeod, "I've seen firefighters with 20 percent of their bodies burned, and *they've* gone home already."

"What you have is more serious," McLeod told her soberly.

Evangeline asked the doctor why she'd gotten the infection. Was it because she was so worn down from having twins, and moving into a new house, that her immune system was compromised? McLeod shook his head. "On the contrary," he said. "This is like getting a pterodactyl through your windshield. Who knows why it happened to you?"

With the worst over, McLeod added, he could tell Evangeline just how lucky she'd been. Had she waited until the morning to call a doctor, she would have died. The bacteria had moved up her arm that night and was half an inch from her lymph system. She was about an hour from dying, McLeod said, when she walked into the Stamford emergency room, and the doctors' measures began, at least, to keep the bacteria at bay.

Evangeline left the hospital on December 18. Her left index finger was badly deformed, but only the tip of it was gone as a result of the infection and Rha's debridements. Evangeline called it her "finger of faith": she was convinced that faith and positive thinking had helped save her life. That and, she felt, her history with cancer. Ever since the cancer, she had taken special measures to keep her immune system up, popping five tablets of wheatgrass in the morning, for example, and five more at night. She felt sure that a strong immune system had played a key role. Her doctors could hardly disagree with her. They knew what necrotizing fasciitis was, and how to throw antibiotics at it, the stronger the better, to try to keep it at bay. But they had no more idea than Evangeline why it afflicted one person and not others, or why, given how long the antibiotics took to kick in, Evangeline had survived her ordeal.

Group A streptococcus, also known as *Streptococcus pyogenes,* of which necrotizing fasciitis was such an obstreperous member, contained about eighty bacterial strains. All lived, initially at least, in the human throat, where they caused infections that ranged from strep throats to scarlet fever and kidney disease. Some streps escaped to the skin in saliva or respiratory droplets. There they brought on, in ascending order of misery, impetigo, erysipelas, cellulitis, and, in the worst cases, necrotizing fasciitis, which tended semantically to include two variants, strep toxic shock syndrome (STSS) and necrotizing myositis.

Sometimes, as in Evangeline Murray's case, necrotizing fasciitis entered through a cut in the skin. Sometimes, to doctors' mystification, it seemed to pass from the throat internally to various parts of the body, then be provoked into infection by a mere bump or bruise. That, apparently, was the case with George Poste, chief scientist for the international drug company SmithKline Beecham. One day in mid-April 1998, Poste had gone to work in perfect health and attended an SKB board meeting. During the meeting, he happened to knock his elbow against a conference table—a "funny bone" knock that hurt, though not too badly. As the day wore on, Poste experienced more pain, rather than less, from his elbow. He also began to feel acutely feverish, though he didn't think to link the fever to his elbow pain. At home that night, he rolled up his sleeve to see that his arm was brilliantly red with radiating streaks. "You're going right to the emergency room," Poste's wife declared. Fortunately, the Postes lived in suburban Philadelphia near Bryn Mawr, a first-class hospital associated with the University of Pennsylvania. As soon as the Bryn Mawr doctors saw the area of intense inflammation advancing up Poste's arm, they pumped him with antibiotics and wheeled him into the operating theater for a surgical debridement. Poste ended up with a small surgical scar on his elbow—and thirty days of continuous antibiotics. Had he waited until the morning, his surgeon assured him, he would have been dead.

Whether by cut or bruise, necrotizing fasciitis often arrived for the start of its feast at an appendage and then began destroying the cells of the surrounding fascial layer of connective tissue beneath the skin and muscle. The bacteria thrived on nutrients there, as well as on fats and muscle above and below the fascial layer. The fascia made for safe grazing, because it had relatively few blood vessels and thus hardly any natural immune defenses at the site of infection. As it destroyed the fascial plane, the bug effectively killed the outer skin by cutting off its circulation. That was when the skin began to turn black and take on a mottled, flaky appearance.

As it ate, often moving up the skin at the rate of an inch an hour, the bug also dispersed exotoxins that killed surrounding tissue. (The toxins might not have been designed to kill, only to break down proteins for the release of nutrients and amino acids the bug wanted to absorb. But in a human host, they were deadly.) As they spread through the bloodstream, the toxins also attacked organs and provoked violent reactions from the immune system. The immune system was designed to attack toxins at a particular site by inflaming capillaries around the toxin to entrap the invader. It *wasn't* designed to take on toxins in all directions. When confronted by the equivalent of a five-alarm fire, the immune system began a destructive inflammatory "cascade" against these superantigens. Specifically, its cytokines—chemical hormones that were part of the normal inflammatory response to any infection—flooded the body in an overreaction, sometimes called a cytokine storm, that caused widespread tissue damage. Even if patients were in the hospital and on the strongest antibiotics at this stage, they had little chance of surviving the infection. The antibiotics killed the bacteria, but the exotoxins were still circulating in the bloodstream, and the immune system's cascade was in full flow, sending blood pressure plummeting and shutting down vital organs—the kidneys, the liver, eventually the heart—in the end game of toxic shock.

All too often, necrotizing fasciitis went unrecognized until this irreversible stage. It was so rare that few emergency room doctors had seen it. Worse, it had no signature symptom when it first appeared.

Victims complained, as Murray and Poste had done, of fever and dizziness and abdominal pain, all symptoms easily mistaken for those of a common flu. The biggest hint was inordinate pain at the site of infection, but that was hard for a doctor to measure; perhaps the patient was a hypochondriac, whining about a bruise that would heal in a day or two. Unfortunately, in a day or two, the patient might be dead.

At least necrotizing fasciitis remained susceptible to most antibiotics—*if* it was diagnosed in time. But why had the flesh-eating bacteria begun to strike in 1989 after decades of quiescence, then spread to more than 1,500 people annually in North America, killing 20 percent of them? Was it just a coincidence that its spike occurred as other Gram-positive pathogens were growing sharply more resistant to various antibiotics? Was it finding some ecological niche created by the ongoing war between bugs and drugs? If not, what accounted for its sudden emergence? And why were a few unlucky people susceptible to it while most others were not?

Dr. Dennis Stevens, an Idaho-based infectious diseases specialist who would sound the alarm on necrotizing fasciitis, called Group A strep "an Old World pathogen." Hippocrates, he liked to point out, had described streplike symptoms in the fifth century B.C. Through succeeding centuries, diseases caused by Group A strep had raged, then retreated, as man and microbe competed for ecological dominance: strains of the bug acquiring new virulence factors and the human immune system learning, eventually, how to combat them. The Group A strep infection scarlet fever had swept the civilized world as a terrifying killer from the 1600s to the 1930s, then gone into a steep decline *before* the advent of antibiotics. The most feared infection of the nineteenth century was "hospital gangrene," another Group A strep. In the Crimean War of the early 1850s, French forces fighting the Russians on the Bosporus were inundated by it, so much so that on one French hospital ship sixty infected men were thrown overboard in thirty-six hours to prevent the whole crew from being infected. Soon after in the Civil War, thousands of Union and

Confederate soldiers lost limbs and lives to the rampaging bacteria. But, like scarlet fever, gangrene waned in the early twentieth century for reasons no one could quite explain.

Antibiotics, along with improved sanitation and hygiene, appeared to further reduce the threat of Group A streps. If *S. pneumo* was the "dumb bug," susceptible to penicillin until the mid-1970s, the Group A streps were dumber. They were *still* susceptible to penicillin and most other antibiotics. Scarlet fever and rheumatic fever were both reduced to minor childhood ailments, and necrotizing fasciitis had all but vanished from sight.

Then, in 1989, Stevens saw a Group A strep infection like no other he'd ever treated or read about. It was a fast-moving infection of the leg that killed an otherwise healthy man despite all the antibiotics that Stevens could throw at it. Stevens realized that the bacteria had produced toxins that led to toxic shock. Perhaps, he reasoned, this was the start of an outbreak similar to the *S. aureus* toxic shock cases of the early 1980s, the ones that had implicated certain brands of tampons. Stevens called Patrick Schlievert, the Minneapolis-based researcher who had solved that mystery—the same Schlievert who would have strong feelings about community MRSA—and the two began sharing ideas about what appeared to be a resurgence of necrotizing fasciitis.

Stevens, a folksy, gentle-mannered fellow whose interest in bacterial infections traced to the deaths of two grandparents from complications of rheumatic fever, started asking colleagues if they'd seen other cases similar to that of his patient with the leg infection. "Oh, sure, I saw one just like that," chorused one after another. Stevens called his network the Rocky Mountain Pus Club, and he began gathering the accounts for publication. He ended up with twenty of them, all from the Rocky Mountain region between 1986 and 1988. The average patient age was thirty-six. These were ordinary people in seemingly good health until their infections flared up. Nineteen of the twenty suffered toxic shock; seven of the group died. The strains of Group A strep varied from patient to patient, but eight of them produced the exceedingly rare pyogenic exotoxin A, which no one had observed in years. As best as Stevens and Schlievert could tell, necrotizing fasciitis appeared to be a new strain of Group A strep with toxins borrowed from some other bacterium, perhaps a long-dormant strain of extremely virulent scarlet fever.

If Stevens and Schlievert's paper in the *New England Journal of Medicine* stirred experts in the field, the shocking death of Muppets creator Jim Henson less than a year later on May 16, 1990, brought Group A strep international attention. Henson, fifty-three, was at the pinnacle of an extraordinary career, his life-size puppets the mainstay of *Sesame Street* and *The Muppet Show.* Earlier that year, Henson had sold his empire to Walt Disney Co. for roughly $200 million. Healthy, happy, and still young, he was about to turn his profits into new creations. One spring weekend, he went to play golf with his brother, a physician. Henson complained of a flu. That evening, in response to his daughter's worried questioning, he said he was "just tired," though added, in an oddly distracted manner uncharacteristic of him, "Hi ho, Kermit the Frog here." On Monday, May 14, he canceled a recording session. In the evening, after coughing up blood, he agreed to take a cab to New York Hospital. By one doctor's account, the cab got stuck in traffic and Henson got out to walk the remaining blocks into the emergency room. By then, as one doctor put it, Henson was a dead man. A tumor larger than a softball was now clogging his lungs, killing tissue around it and generating lethal toxins that brought on toxic shock. Technically, he succumbed to one of the two Group A strep strains of necrotizing fasciitis: strep toxic shock syndrome (STSS). In a loose sense, what killed him was necrotizing fasciitis of the lungs.

Up in Ontario, Canada, Don Low read the reports of necrotizing fasciitis with keen interest. He began to suspect that it had appeared as the result of a waning of "herd immunity" against scarlet fever or some other Group A strep. If the theory was true, an earlier generation had acquired antibodies to the dreaded infection after widespread public exposure to it. Eventually, almost everyone developed immunity to it. As the disease seemed to vanish for lack of susceptible hosts, a new generation of people had no need to produce antibodies to it, so their bodies grew without them. Then, perhaps, a new serotype[1] of Group A strep acquired some of the original infection's virulence factors, including its toxins, which enabled it to cause and spread disease to now vulnerable hosts.

Why the bug infected certain people and not others was a mystery. Clearly some host factor was involved, or else all of the 20 percent of healthy people who carried Group A strep infections in their throats at any given time would get necrotizing fasciitis. Low noticed a correlation between the incidence of disease and the weather: the colder the climate, the higher the incidence, especially in winter. Possible risk factors appeared to include the human immunodeficiency virus (HIV), cancer, diabetes, alcohol abuse, and chickenpox (the bacteria could enter an immunocompromised bloodstream through open lesions, or so went the theory). Whatever the real list was, chances were that George Poste, Jim Henson, Evangeline Murray, and other victims had shared one or more of them.

On the coffee table of his hospital office, Low, a round-faced, bespectacled fellow of about fifty who dressed in plain oxford shirts and blue jeans and had a lack of professional pretense to match, spread color photographs of some of the more gruesome necrotizing fasciitis cases he began to see as the 1990s unfolded. Here, he pointed out, was a man with his chest filleted as if by a psychopathic killer, a woman whose leg had turned purplish black, the dead skin hanging in flaccid folds: a dozen or more 5x7s, each more horrifying than the last. Low's cases had mounted to fifty in 1992. Then, oddly, the bug seemed to vanish. Just as Low and his colleague, Allison McGeer, were about to consign the outbreak to history, a new epidemic occurred. This time, the future premier of Quebec, Lucien Bouchard, was among the victims.

One Monday in the fall of 1994, Bouchard was admitted to Montreal's Saint-Luc Hospital with pain in his leg that his hematologist thought might be a blood clot. Low was called in to consult. On

---

[1]Serotype is a rank of classification below the species level based on an organism's reaction with antibodies in serum.

Wednesday morning, the combination of symptoms set off an alarm. Tests revealed that the infection was a necrotizing myositis—the other variant of necrotizing fasciitis—that had destroyed much of the fascia in Bouchard's leg and begun munching up toward his abdomen. Low urged the attending doctors to amputate Bouchard's leg at mid-thigh and to put him on a heavy course of immunogammaglobulin, or IVIG. Bouchard's amazing recovery in twenty-four hours showed that the pooled antibodies could neutralize the toxins and, as Low had hoped, keep the immune system from being overstimulated, though to survive, he did have to lose part of his leg.

Out in Boise, Idaho, Dennis Stevens observed the regular outcroppings of flesh-eating bacteria of the 1990s with increasing curiosity. He felt that by then most of the population should have developed a herd immunity to it. After all, as a Group A strep, necrotizing fasciitis was also associated with milder and far more widespread conditions, such as run-of-the-mill sore throats. As those symptoms pervaded the country, they should have provoked antibodies that would also combat necrotizing fasciitis. That they didn't was probably because there was more variety within the bacteria's own population than any one kind of antibody could handle. IVIG was therefore a sensible treatment, but Stevens knew of patients who had had near-fatal anaphylactic shock reactions to it and others who had shown no reaction, good or bad, to the stuff. Stevens thought there was no substitute for early use of antibiotics, particularly clindamycin, and surgical debridement.

Stevens did a study of his own that suggested the bug had picked up a particular toxin in 1985, one that appeared in 99 percent of all cases after that time and in none of the cases before. But this told Stevens nothing about who was susceptible or what the host factor was. Geographically, for example, it seemed to strike at random. Stevens tracked a late-1990s miniepidemic of cases in the Rocky Mountain area, including nine in less than a year in Missoula, Montana, a figure far higher than the CDC's national estimate of cases per hundred thousand population. While nine was still a relatively low number, nearly half of those people died, and most of the others lost arms or legs. Why Missoula? Who knew? Mostly, the bug cropped up in the community, appearing one month in Minnesota, the next in California, but now and again it hit hospital patients, too.

Nor was there any socioeconomic, ethnic, gender, or age pattern. In Queens, New York, an eight-year-old boy died of it, apparently as a consequence of chickenpox. In Chicago, a seventy-year-old man was one of fifteen deaths in the area in 1999. That same year, a sixty-one-year-old Cuban American man was mowing his lawn in Florida when he noticed that a small cut on his arm had grown oddly inflamed; he died soon after. In San Francisco, also in 1999, two prostitutes died of flesh-eating bacteria after sharing needles to shoot up with black tar heroin. "When a person is admitted to the intensive care unit and within hours is dead," Dr. Thomas Aragon, director of community health epidemiology and disease control at the San Francisco Health Deparment, said soberly after watching the prostitutes expire in great pain, "you develop a real appreciation for bacteria and what their toxins can do."

Vincent Fischetti, a well-known Ph.D. specialist with thirty-five years of experience in Group A strep infections at New York's Rockefeller University, sat in his peaceful, sun-filled office one day and blithely predicted the worst-case scenario. "Obviously necrotizing fasciitis *will* become resistant to penicillin," he mused. "It's just a matter of time. These things take a number of hits for an event to occur. This is probably a rare event, and not enough hits have occurred. But with more organisms becoming resistant—some more related to strep than they ever were before—the chances are greater than ever. It will happen, and when it happens, it will be a very scary event."

Richard Novick, one of the most distinguished researchers in bacterial pathogens of his time, begged to disagree. In his own office twenty-five blocks south of Fischetti's at New York University's Skirball Institute, Novick observed that Group A strep *should* have acquired penicillin resistance long ago. The classic beta-lactamase enzyme that conferred resistance by hydrolyzing penicillin was very active, he observed; it ought to have made its way to Group A strep. In fact, staph bacteria that did have the enzyme were often in the throat, right along with Group A strep—so commingled that when penicillin was given for a strep throat it often failed to work because the resident staph zapped it with beta-lactamase, like a soldier from one regiment protecting an unarmed soldier from another. Yet the enzyme never jumped over. Why? "One suspects that the beta-lactamase enzyme is probably detrimental to the streptococci," Novick theorized. Still, wasn't it just a matter of time, as Fischetti suggested, before the strep figured out how to accommodate it? "I don't think so," Novick said. "Penicillin has been around for fifty years. They used to treat every case of streptococcal rheumatic fever that you could think of with it." Indeed, widespread use of antibiotics had reduced the incidence of rheumatic fever dramatically. "And all that time, *S. aureus,* the most common organism with beta-lactamase there is, was in the same environment. Those beta-lactamases can be transferred. But they never do get transferred to Group A strep. So I don't think it's going to happen."

Perhaps Novick was right. But how disconcerting, in a time of such microbiological sophistication, that no one really knew if Group A strep infections, including those that cause necrotizing fasciitis, would grow resistant to penicillin next month—or perhaps stay susceptible for the next fifty years. The experts could theorize and they could debate, but in the end all they could really do was wait to see what happened. In truth, they weren't in control: the flesh-eating bacteria were.

Possibly as a harbinger of trouble to come, researchers in Pittsburgh reported, in the spring of 2002, an unprecedented spike in the United States of Group A strep resistant to erythromycin. Erythromycin is in the class of antibiotics known as macrolides, not the beta-lactams, which include penicillin. Nevertheless, the news was alarming. For the 5–10 percent of Group A strep patients allergic to penicillin, macrolides were the next drugs of choice. The strain identified by Dr. Judith M. Martin and colleagues at Pittsburgh's Children's Hospital was resistant, they noted, not only to erythromycin but to all the macrolides, including a hugely popular drug, azithromycin (sold under the brand name Zithromax). In 1998, when the team's three-year study of children between kindergarten and eighth grade began, erythromycin resistance in Group A strep had been noted in the United States, but generally at rates of 2–3 percent in whatever group was tested. Fifteen percent of the Pittsburgh children at that time tested positive for strep throat, but all their strains were susceptible to erythromycin. In the third year of the study, Martin found that 18 percent of the children she studied had strep throat. Shockingly, of that 18 percent, 48 percent had erythromycin-resistant strains. Resistance had also spread to 38 percent of grown-ups in the community, she found. For years, Martin said upon the study's publication in the *New England Journal of Medicine,* doctors had tried to ignore spiking rates of macrolide resistance in several European countries and Japan. Now the bug had come west. "We've talked about this for years," Martin said, "and now it's here. We can't sit back and say, 'Not us.' It's clearly here."

Did Group A strep's newfound resourcefulness with macrolides suggest that it would soon master the genetic trick of penicillin resistance? Or did the two have nothing in common? No one could say for sure. If worse came to worst, and streptococci that caused necrotizing fasciitis did become resistant to penicillin, the death toll, at least, would be relatively small, painful as those deaths would be. The prospect of *S. aureus* acquiring vancomycin resistance from enterococci remained far scarier. But to many doctors in clinical practice, a truly alarming rise in resistance had already occurred, one that had nothing to do with *S. aureus* or streptococci or any other Gram-positive pathogens.

After three decades of quiescence, the Gram negatives were getting resistant, too.

12

MORE BAD NEWS

For the better part of a decade, a quickening drumbeat of concern had been sounded about resistant Gram-positive pathogens. Finally, its intended audience had begun to hear. In America's communities, doctors often hesitated now to prescribe antibiotics for every cough and earache, and many parents understood why. Hospitals set tighter rules on which drugs could be used without approval. And many big pharmaceutical companies had drugs for Gram-positive organisms in at least some early stage of development. Now, as pressure mounted on the Gram positives, an utterly predictable, Darwinian exercise in natural selection was occurring.

Experts in the ecology of bacteria called it the squeeze-the-balloon phenomenon. Squeeze one end of a sausage-shaped balloon and the other side bulged. In some hospitals, Gram negatives actually posed a greater threat by the winter of 2002 than Gram positives. Some Gram negative strains were *pan*resistant—resistant to every drug, the ultimate medical nightmare. Vancomycin-resistant enterococci had been panresistant, too, before the advent of Synercid and linezolid, but VRE was a second-rate pathogen. Gram negatives killed. And there was no new drug in the pipeline for them—none.

Gram negatives preyed on very ill hosts, which meant they resided in hospitals for the most part, moving among elderly or chronically sick patients. But a young, previously healthy person subjected to trauma—from a burn, a car crash or other accident, chemotherapy, or an organ transplant—was an opportunity for Gram negatives, just as he was for Gram positives. Three Gram-negative pathogens— *Pseudomonas aeruginosa, Klebsiella pneumoniae,* and *Acinetobacter baumannii*—had become as infamous in hospitals at the dawn of the twenty-first century as the Big Three Gram positives had over the previous decade. Of the 2 million cases of infectious diseases said to involve Americans each year, the CDC estimated that 40 percent were now caused by Gram-negative pathogens.

Under a microscope, many Gram negatives look like little rods, though some are also grapelike "cocci." Each Gram-negative cell has a two-layer membrane, the outer of which has little channels, called porins, through which nutrients can pass. When certain Gram-negative cells are hit with antibiotics, they can alter those channels and become resistant to the action of the incoming antibiotic—a mechanism that Gram positives lack. This outer membrane also carries some nasty virulence factors, including toxins that can cause an overreaction by the human immune system, which can lead to sepsis and death.

Like Gram positives, Gram negatives are human commensals: benign flora among the billions of bacteria carried by every healthy human being. *Klebsiella* is mostly a resident of the intestinal tract; *Acinetobacter* and *Pseudomonas* tend to colonize the respiratory tract. But given the right chance in an immunocompromised host, all cause a witch's brew of trouble. From the GI tract, *Klebsiella* causes infections in the urinary tract, the gall bladder, the bile ducts, kidneys, and more, leaving a path of destruction in its wake. *Pseudomonas* can do the same. Both also love nothing more than a puncture wound that spills them directly into the bloodstream, where they can cause peritonitis (infection of the abdominal fluid) or bloodstream infections. *Acinetobacter* is apt to cause pneumonia. But, like terrorists attacking from all directions and with every kind of weapon, the Gram negatives defy easy categorization. *Acinetobacter,* once a relatively harmless bug, causes bloodstream infections; *Klebsiella* causes pneumonia. All three, indeed, can infect any organ of the human body.

In part, this ability is due to their ubiquity. *Acinetobacter* likes the respiratory tract but also resides on moist areas of the skin. *Klebsiella* is partial to the GI tract but, like *Acinetobacter* and *Pseudomonas,* can find its way to the lungs. Worse, Gram negatives aren't simply residents of the human body, inside and out. They're *everywhere.* Somehow, long ago, many Gram positives worked their way up from the soil to animals and humans, and there, for the most part, they remain. *S. pneumo* is found now only in the human throat; *S. aureus* is mostly on human skin or the nostrils; *E. faecium* moves from the gastrointestinal tracts of animals to humans and populates the feces of both. Gram negatives *still* reside in soil, with or without animals or humans around. Many like water (especially *Acineto-bacter,* known as a "water bug.") They *love* vegetables: a thick, invisible layer of *Pseudomonas* is what causes vegetables to rot at room temperature. They *adore* flowering plants, so much so that plants are prohibited from the rooms of patients on chemotherapy, whose ravaged immune systems are especially susceptible to flower- or plant-borne *Pseudomonas.* And, like some Gram positives, many Gram negatives cling tenaciously to all surfaces—sheets, bedrails, soap dishes, stethoscopes, ventilator tubes, catheters, and the hands of doctors and nurses, to name just a few—from which they're perfectly capable of alighting onto their next human victim and causing infection again.

Once upon a time, before antibiotics, Gram negatives played relatively minor roles in human misery. Certainly they caused infections and death—Friedländer's pneumonia, a kind of *Klebsiella,* was defined in the nineteenth century—but because they preyed strictly on long-dwindling, compromised patients, they rarely had the chance to knock off a patient before something more virulent did. Not until the 1940s was the first *Pseudomonas* infection even described in medical literature.

Penicillin, it turned out, had no clinical effect on many Gram negatives. This was partly because the drug and the bugs both had negative charges, so their electrostatic reactions repelled each other, and partly because the bugs had beta-lactamases. And it was also because they had porins they could alter in order to keep beta-lactam drugs from diffusing into their cells. Penicillin actually fostered the *spread* of Gram negatives by clearing patients of their Gram-positive infections, leaving the field open to Gram negatives. As chronically ill hospital patients began to live longer, they became ever tastier fodder for Gram negatives, especially with the advent of two more life-extending innovations: dialysis machines and ventilators. The tubes and catheters by which chronically ill patients were tethered to these machines served as so many footbridges for billions of the once ignored bugs, all drawn by the lure of critically vulnerable human hosts.

Eventually pharmaceutical companies awoke to the threat, first coming out with amoxicillin and ampicillin, then cephalosporins, then second-generation cephalosporins. Each time, the bugs retaliated with new beta-lactamases, which they proceeded to trade among themselves. Increasingly, the war of one-upmanship between bugs and drugs came to seem like the old *Mad* magazine cartoon "Spy vs. Spy." A third-generation cephalosporin called ceftazidime seemed to stymie the Gram negatives at last—until 1983, when to fight this extended-spectrum beta-lactam drug the bugs produced yet another mutant enzyme that cut the drugs' chemical rings. Crestfallen scientists called these, logically enough, extended-spectrum beta-lactamases, or simply ESBLs. Now various strains of the Gram negatives share more than 150 plasmid-mediated ESBLs, enough to knock out any new cephalosporin, on its own or in combination with any other drug, that big pharma can dream up.

It was in the midst of that global evolution among Gram negatives that Dr. James Rahal became the head of infectious diseases at a hospital in Queens, New York, and found himself immersed in the worst infestation of ESBL Gram negatives documented in any hospital in the United States.

James Rahal is a gentle figure of modest height and professorial mien, blessed with the patience to elucidate complex science in simple terms, for hours at a time, to laypeople. He is now regarded as a superb infectious disease clinician of national stature, but in 1988, when he took over the infectious disease department at Booth Memorial Hospital on Main Street, Queens, he was seen, if anyone noticed him at all, as something akin to the manager of a farm-league baseball team.

Booth Memorial, now known as New York Hospital Queens, was and is a 500-bed community institution populated by mostly elderly, impecunious patients passed like bad pennies from third-rate nursing homes. Bacterially, as one well-known East Coast expert in the field put it, Rahal's hospital, like many in New York, was ripe with the usual suspects. But Rahal had two advantages starting out that many ID chiefs in larger hospitals lack. For one, he really knew his microbes. A former fellow of his had developed a software program called Gideon for diagnosing infections: you put in half a dozen variables and the computer spat out five microbial suspects, from the least likely to the most likely organism, based on mathematical algorithms. Rahal could do the algorithms in his head and come up with the same list of suspects as the computer every time. Because of that, and because of his empathetic manner, Rahal also had an administration that believed in him, as well as a fiercely loyal staff.

Early on, Rahal realized that a number of the hospital's patients had similar strains of *Acinetobacter*. The strains were susceptible to ceftazidime and a stronger drug, imipenem, but resistant to everything else. Rahal decided to blitz the *Acinetobacter* with ceftazidime. Unfortunately, he began seeing resistance to the drug, not only in *Acinetobacter* but also in *Klebsiella*. Within two years, Rahal had an epidemic of ceftazidime-resistant *Klebsiella*.

In desperation, Rahal resorted to imipenem, a drug he liked to call the vancomycin of Gram negative antibiotics. In most cases, he found, imipenem worked. It knocked out ceftazidime-resistant *Klebsiella,* and it knocked out ceftazidime-resistant *Acinetobacter*. Then one day Rahal's lab chief stuck his head in Rahal's office with an agar plate and a worried expression. He'd just seen his first *Klebsiella* resistant to imipenem, and he knew just how chilling the implications were. With imipenem useless, a number of infected patients began to die because their strains were resistant to all clinically available drugs.

That was when, with his hospital's blessing, Rahal took Draconian action. Resistance to ceftazidime was what had enabled the bugs to become resistant to imipenem. It was a stair-step process. By studying formulary records, Rahal realized that ceftazidime use at his hospital had increased 600 percent during the two years preceding the outbreak. Clearly, the more antibiotics one threw at the bugs, the more resistance they acquired. That had to change.

The answer was to stop using ceftazidime and all other cephalosporins. Then perhaps the ceftazidime ESBLs would go away. Once the ceftazidime ESBLs were gone, Rahal could hope that resistance to imipenem might melt away, too. So no doctor in the hospital would be allowed to prescribe cephalosporins without Rahal's permission, and permission would be granted only if a patient's life was imperiled and no other drug worked. Many patients over the next three years probably took a little longer to get well on slower-acting drugs, but no one died, and in the surgical intensive care unit, where cephalosporins were most prevalent, Rahal's policy reduced the incidence of ceftazidime-resistant *Klebsiella* by 87.5 percent.

Within a year, nearly all the hospital's *Klebsiella* were susceptible again to all cephalosporins, including ceftazidime, and, more important, to imipenem. Unfortunately, the campaign had created a new squeeze-the-balloon problem. Using all that imipenem and holding ceftazidime in reserve had brought *Klebsiella* and *Acinetobacter* down to manageable levels. At the same time, the policy had spurred an upsurge of imipenem resistance in *Pseudomonas aeruginosa*.

In some ways, *Pseudomonas* was the worst of the three most prevalent resistant Gram negatives. Water-based products were in constant use for changing skin dressings for burn patients. Burn surgeons came to dread its fruity, sweet smell when it infected the skin, and its bluish green tinge. It was a prominent cause of pneumonia, both in and out of the burn unit, a complication of lung transplants and a leading cause of death in cystic fibrosis patients. It caused urinary tract infections and led to septicemia. Most troubling, it usually acquired resistance to a new drug more quickly than *Klebsiella* and *Acinetobacter*.[1]

Mercifully, many of these *Pseudomonas* strains had acquired their resistance to imipenem so directly that they were still susceptible to cephalosporins. Rahal had to mete out the cephs carefully to avoid triggering resistance there, too. In most cases, he could do that. What made him sweat were the exceptions: *Pseudomonas* resistant to imipenem *and* cephalosporins. Sometimes, all he had to fall back on was a pharmaceutical nightmare called polymyxin.

Polymyxin was, as Fred Tenover of the CDC freely put it, a "horrible" drug. It killed a lot of pathogens, but it killed people, too. Some 25 percent of patients on polymyxin developed kidney failure; in many others, it induced internal kidney ulcers. Not surprisingly, it had been a drug of last resort for more than two decades, used so rarely that most doctors considered it obsolete. But for some of the multidrug-resistant *Pseudomonas* that Rahal was seeing, polymyxin was all he had left. Knowing just how toxic it was, he prescribed it cautiously and hoped for the best. By and large, it helped keep *Pseudomonas* in check—for the moment.

By 1998, Rahal could report that his epidemic of ceftazidime-resistant *Klebsiella* was over. The imipenem resistance he'd provoked in *Pseudomonas* was down, too. Not gone, but down. Sometimes, in his windowless basement office, he felt like one of those old comedians on *The Ed Sullivan Show,* spinning plates on sticks. One plate would lose momentum and threaten to fall, and Rahal would race over to spin it again. As he did, another of the plates would slow and start to wobble. One Gram negative would grow resistant, so Rahal would try a new drug against it. Maybe the new drug worked, maybe it didn't; if it did, it almost certainly provoked resistance in *another*

[1]So quick was it to acquire resistance that in Rahal's hospital it had grown resistant to imipenem without the stair-step path of getting resistant to cephalosporins first. It didn't need to import ESBLs from some other bacteria or develop them from chance mutations. It could close its membrane channels—the channels called porins—and simply block imipenem from reaching its target site.

bug. All Rahal could do was keep balancing the drugs and hope for the best. For now, the best was good enough.

Over in Brooklyn, a short drive from Rahal's hospital through a crazy quilt of ethnic neighborhoods, David Landman, M.D., began wondering, in the late 1990s, if he could do anywhere near as well. Landman was on the infectious diseases team of the hospital known as SUNY Downstate: State University of New York Health Science Center. By 1997, Landman and his colleagues were so alarmed by the rise of all kinds of resistant Gram negatives at hospitals throughout Brooklyn that they initiated a boroughwide study. They found that 44 percent of the *Klebsiella pneumo* isolates they tested from fifteen Brooklyn hospitals had ESBLs that made them resistant to cephalosporins, if not also other drugs. Most of the isolates were from a few strains, a sure sign that the bugs, once provoked to resistance by cephalosporins, were spreading on the hands of healthcare workers and other environmental surfaces, as well as being spread by the patients themselves as they were shuttled from hospital to nursing home and back. The level of resistance, they reported soberly, was endemic.

Acinetobacter's story was particularly chilling. At those same fifteen hospitals, only half of the isolates were susceptible to imipenem and the other carbapenems. Only 25 percent were susceptible to ceftazidime and the cephalosporins. Landman could see a crisis unfolding. Why was it happening in New York? he wondered. And what, aside from reaching for the dangerously toxic polymyxin, could be done about it?

The epidemic did seem to be targeting New York, though reports from elsewhere ran the gamut. Reports of epidemic Gram-negative resistance as the 1990s unfolded came from France, from Spain, from Japan. In Zurich, Switzerland, two doctors studied six types of ICUs in European tertiary-care medical units between 1992 and 1999. They found that imipenem resistance in *Pseudomonas aeruginosa* rose from almost nothing to 12 percent between 1992 and 1997. Worse, between 1998 and 1999 it rose to 30.9 percent. That was an astonishing increase in just one year.

In the United States, some hospitals reported 40 percent resistance rates among *Klebsiella* isolates, for example, to many antibiotics, while others encountered no resistance at all. Such disparities characterized the other Gram negatives, too. In the fall of 2001, at the Cleveland, Ohio, VA Medical Center, Louis Rice, M.D., felt relatively sanguine about most of the Gram negatives he was seeing in his city hospitals but had begun to worry about new strains of multidrug-resistant *Acinetobacter*. At Brown University's Miriam Hospital in Providence, Rhode Island, Antone Medeiros, M.D., reported a sharp rise in resistant *Pseudomonas* rather than *Acinetobacter*. At the Lahey Clinic in Burlington, Massachusetts, George Jacoby, M.D., reported strains of *Acinetobacter* that were widespread but relatively benign—like crab grass, as he put it. The bugs' variability was hardly reassuring: it meant that a mild threat on one day might be a ferocious one the next.

Generally in the United States the worst resistance rates occurred on the East Coast. At the Baltimore VA medical center where Glenn Morris worked, Judith Johnson of the hospital's infectious disease staff called the rise of ESBL Gram negatives "truly terrifying." A wry, no-nonsense microbiologist, she noted that "what you see historically with any resistance problem is that at first you get a few isolated cases and someone publishes a study. Then five years later, you might see .01 percent, and five years later you see 5 percent, and then two years after that you see 50 percent, and you can no longer use the drugs against that organism." Susceptibility rates were still high with Gram negatives in most cases, she explained, but resistance was rising enough that doctors could no longer count on certain drugs—the cephalosporins, for example—as empiric therapy. "That's one of the things that gets harder and harder to figure out, what kind of regimen you can use before you know what the patient has."

From his bunker in Queens, Jim Rahal often mused on why New York in particular seemed so hard hit by resistant Gram negatives. He could drive across the Hudson River and find hospitals in New Jersey that had no resistant Gram negatives at all. Why was that? One reason, surely, was socioeconomic. New York Hospital Queens served generally poor, elderly people. The New Jersey hospitals he'd seen catered to more affluent patients—better fed, presumably, with life-times of better medical care. Often, too, his patients came from nurs-ing homes, where resistant bugs spread among an ideal population of immunocompromised hosts. And even the top private hospitals in New York were probably affected disproportionately by resistant bugs because New York was a major referral center. Patients afflicted with cancer and other chronic conditions came from all over the northeast to be treated in city hospitals. These were patients who got a lot of antibiotics; they were also perfect hosts for Gram negatives.

At his own hospital in Brooklyn, David Landman did manage to beat back ceftazidime-resistant *Klebsiella*—Rahal's early nemesis— by using a drug called piperacillin/tazobactam. For the pip/tazo-resistant *Acinetobacter* that he got as a result, he had other drugs to use—including imipenem, whose usefulness he'd reserved by not re-sorting to it as his principal fallback when ceftazidime failed. Still, he began to feel he was fighting a losing battle. As patients flowed in from other Brooklyn hospitals and regional nursing homes, they brought imipenem-resistant *Acinetobacter* with them. Increasingly, he saw iso-lates resistant to imipenem and the cephalosporins (like ceftazidime) and all the other antibiotics in common hospital use. By the fall of 2001, he was seeing isolates resistant to even the dreaded polymyxin. That was appalling, because it meant that for those strains Landman had nothing left to give.

Just how easily could these newly multidrug-resistant *Acinetobacter* strains travel in such wards? At nearby Kingsbrook Jewish Medical Center, a Ph.D. colleague of Landman's named Steven Brooks con-ducted a chilling experiment. He'd noticed that at his own hospital strict enforcement of the established infection control standards for *Acinetobacter* often failed to stop the bug's spread. That prompted a theory. What if it was being spread not just by contact but by airborne transmission?

Brooks took petri dishes filled with an agar medium that supported bacterial growth and placed them at measured distances from the headboards of patients with *Acinetobacter* infections of the respiratory tract. The dishes were spaced out to every wall of each patient's room and even out the door. When the dishes were analyzed some hours later, Brooks found that about 40 percent of the dishes set one foot away from the patients had active *Acinetobacter*. But so did the same percentage of dishes set eleven feet away. "In several instances," Brooks reported in a letter of May 2000 to a medical journal, "*Acinetobacter* was also detected on sedimentation plates placed outside of the patient's room and as far away as the nursing station—approximately 22 feet from the room." The strains were identical to those found in the patients' lungs.

Brooks was careful to note that the experiment did not prove *Acinetobacter* could colonize or infect people at those distances, only that the bugs could spread that far. But at his own hospital, precautions were upgraded from "contact" to "droplet" standards. All healthcare workers were required to wear surgical masks and eye protection within three feet of a patient with respiratory *Acinetobacter*.

As far as Brooks was aware, no other hospital in the United States, or elsewhere, for that matter, had yet followed suit.

Acinetobacter, Klebsiella, and *Pseudomonas* were the big three: the Gram negatives with growing resistance that bedeviled hospitals in the United States and abroad, prompting studies and speeches, warnings and fear. But if they were the most widespread, other Gram negatives cropping up in isolated instances provoked as much concern, if not more.

E. coli was the enteric pathogen that traveled easily from the guts of animals into the guts of humans, in all likelihood conferring antibiotic resistance as it went. It was, in fact, a Gram negative. At least the most virulent kind, *E. coli* 0157:H7, had not acquired resistance to antibiotics. The same could not be said, however, of a new *E. coli* strain causing severe urinary tract infections in women in three U.S. cities.

In a study published in the fall of 2001, Amee Manges and colleagues from the University of California at Berkeley found that 55 of 255 female university students on their campus with persistent UTIs, as urinary tract infections are known, had *E. coli* resistant to trimethoprim-sulfamethoxazole, a drug commonly used to treat the condition. Tests showed that a majority of the resistant infections were caused by a single strain, suggesting a recent emergence and rapid spread. The strain was a nasty one, resistant to several antibiotics and imbued with virulence factors that increased its ability to infect the bladder. When Manges et al. compared the Berkeley isolates to those of university women with UTIs in Ann Arbor, Michigan, and Minneapolis, Minnesota, they were astounded to find the same exact strain in 38 percent of the Michigan cases and 39 percent of the Minnesota cases. The strain was still susceptible to other antibiotics, but marked a worrisome trend. If it could spread so quickly around the country—possibly through sexual intercourse, though that remained unproven—what might keep it from soon becoming a pervasive, panresistant pathogen?

Far rarer, but very alarming, were the reports of a drug-resistant form of one of humanity's deadliest enemies—bubonic plague.

In the waning days of empire for France, the Pasteur Institute had established stations in various far-flung colonies as a way to track the global spread of diseases, and, though most of these colonies had become independent, the stations remained in force. One was on the island of Madagascar, off eastern Africa. Here, in 1995, Marc Galimand, a colleague of Patrice Courvalin's, was called in reference to the case of a sixteen-year-old boy who appeared to have malaria. The boy, from the island's Ambalavao district, presented with fever, chills, and muscle pain, all common malarial symptoms, and so was treated with quinine. Three days later, his fever spiked sharply, he became delirious, and near his groin grew a pus-filled bulge, or bubo: the classic sign of *Yersinia pestis,* better known as bubonic plague.

The boy's bubo was punctured and drained, and he was treated with first one, then another, of the drugs proposed for plague treatment by the World Health Organization. To the doctors' surprise, the plague persisted. Not until they tried intramuscular injections of streptomycin and oral trimethoprim-sulfamethoxazole did his infection retreat. When Galimand tested cultures from the boy at the Pasteur's local lab, he was astounded to find that they were resistant to *all* the drugs proposed by the WHO for plague treatment. What was worse, the bug's resistance was plasmid mediated. In one step, by conjugation, this strain of *Y. pestis* had acquired resistance to all those drugs

The implications of Galimand's finding were terrifying. In the roughly 2,000 years since it had appeared as a human pathogen, plague had caused scores of pandemics around the world, almost always in crowded urban settings, killing an estimated 200 million people. It began as a bacterium in the soil that colonized the guts of fleas, often passing from one flea generation to the next without incident. For some reason not yet understood, the bacterium could stir suddenly to action, blocking the passage of food to the flea's gut and causing the flea, in effect, to starve. As the flea grew desperately hungry, it developed a voracious thirst for blood that might sustain it. That led it to the nearest warm-blooded rodent, which often turned out to be a rat. There, nestled in the rat's warm hair, the flea bit in for a tasty meal and so infected its new host. When the rat drew close to humans, foraging for its own meals among human food stores, the flea leapt up for its own next meal, of human blood, and so spread the infection again. Over untold eons, *Y. pestis* had acquired a cluster of genes that made it devastatingly pathogenic once it entered the human bloodstream. Usually the bug first infected the lymph nodes, causing them to swell and forcing pus-filled "buboes" up from the skin, often around the groin and armpits. If the infection progressed into the bloodstream, it destroyed internal organs, caused hemorrhaging of blood vessels, bleeding from the nose and ears, and often brought on dementia. In a small percentage of cases, plague took over the lungs and caused an infectious pneumonia. Then the hapless carrier began to transmit it by coughing it in blood droplets, dramatically accelerating its spread. In days, entire cities would be decimated. This, of course, was devastation due to susceptible strains, before the age of antibiotics.

Streptomycin, chloramphenicol, and tetracycline had reduced the global threat to a seeming historical footnote by the 1960s, but serious students of plague, knowing of the bug's capacity to lie dormant in soil for decades until stirred up by some change in nature, never felt complacent. In India in 1994, an earthquake apparently provided just such a change, churning up buried bacteria and starting the chain of infection from flea to rat to human. After widespread panic and a cost of about $2 billion, much of it in lost tourism revenues, the toll of infection was 6,300 people, but only 50 fatalities—principally because India's revived strain of Y. pestis was entirely susceptible to antibiotics.

So alarming was the Madagascar strain in its implications that for some time Patrice Courvalin had trouble persuading his own colleagues to send it to him in Paris for further study. When he did get it at last, he and Galimand confirmed that the strain was resistant not only to chloramphenicol, streptomycin, and tetracycline but to the best alternatives: ampicillin, kanamycin, spectinomycin, and minocycline. It did remain susceptible to cephalosporins, certain aminoglycosides, quinolones, and trimethoprim, a somewhat reassuring array. But disturbing questions remained unanswered. Was the strain isolated or widespread? Had the bug acquired its multidrug-resistance genes from the boy's own gastrointestinal tract, or were multidrug-resistant strains of Y. pestis circulating among the area's fleas and rodents? Most pressing, was this a natural strain or had it been engineered by some rogue state for bioterrorism?

For some time, the CDC's Fred Tenover believed that the Madagascar strain *was* engineered. As a natural mutation it was just so *unlikely,* given the range of antibiotics to which it was resistant. (Not all of those antibiotics, that is, were given as standard therapy in Madagascar or, for that matter, in East Africa. Why would the bug have become resistant to them?) Tenover worried about who might have engineered the strain. He also worried about how well the Pasteur Institute was safeguarding it. At one point, Courvalin received a request from the CDC asking that he send a portion of the strain to Atlanta for analysis. Courvalin refused. "That was the right answer," he was told. "We just wanted to check to see if you were willing to disseminate the strain."

Eventually, Tenover came to feel the strain was naturally occurring after all. But while that eased one concern, it raised another. Naturally occurring strains, as it happens, are easier to "weaponize."

In the fall of 2001, with the United States still reverberating from the attacks on the World Trade Center and the Pentagon, *Y. pestis* ranked prominently on a list of pathogens said by the FBI to be manipulatable by terrorists for widespread biological warfare.[2] As a weapon of terrorism, a strain of bubonic plague would be ideal: the kind that could be disseminated by aerosol in airborne droplets, then be spread rapidly by coughing from one victim to the next. Several days after breathing in the droplets, the next victim would become infected. With failure to seek prompt treatment, the victim would slip precipitously into kidney and respiratory failure, then into lethal shock. The number of people who could be killed by such means would be limited by rapid use of antibiotics—unless, of course, the strain being disseminated was resistant to antibiotics.

Turning drug-resistant plague into a weapon of bioterrorism was a complex process, almost certainly requiring the sustained effort of an enemy state. But according to Dr. Ken Alibek, former deputy chief of the Soviets' germ warfare agency, Biopreparat, this effort had already been undertaken and had achieved success. Drug-resistant plague, he confirmed upon his defection to the West in the early 1990s, had been among the resistant pathogens manufactured by a network of forty facilities in fifteen Soviet cities with a workforce that numbered in the tens of thousands. Even when the Soviet Union was teetering on the verge of financial collapse in 1990, the annual budget of Biopreparat had been almost $1 billion. The scientists had not been satisfied merely to collect naturally resistant strains; they had used genetic engineering to create new ones. At one point, Alibek confirmed, Biopreparat had maintained twenty tons of plague. Alibek felt sure that some of it remained intact, though stores might only remain virulent for a few years.

If a rogue state proved unable to buy potent drug-resistant *Y. pestis* on the black market, it could always try hiring the scientists who had concocted the stuff in the first place. Alternatively, a terrorist could

---

[2]The three other most worrisome pathogens are anthrax, botulism, and smallpox.

travel to current-day outbreaks of plague and take cultures of the bug back to the laboratory. Globally, the CDC reported, about 3,000 cases still occurred each year. All outbreaks were duly reported as they appeared—information readily available to any resourceful terrorist. In October 2001, for example, an outbreak occurred in Uganda, killing fourteen villagers in three weeks.

Four years after their initial finding, Galimand, Courvalin, and other colleagues at the Pasteur reported a second case of multidrug-resistant *Y. pestis* on Madagascar. "The resistance genes," the team reported, "were carried by a plasmid that could conjugate at high frequencies to other *Y. pestis* isolates." So this, too, was a naturally occurring case, or so the experts thought. That was a relief. But at the rate bacteria multiplied, no microbiologist would be startled to learn, in a decade's time, that this same strain of multidrug-resistant bubonic plague had spread over Madagascar and beyond—unless a band of terrorists got hold of an isolate, in which case Madagascar's multidrug-resistant *Y. pestis* could be spread a whole lot faster than that.

Even without these chilling implications, the prospect of Gram negatives rising in resistance—from the ubiquitous if relatively mild *E. coli* to the possibly unique but horrifying *Y. pestis*—casts a long shadow over a landscape already darkened by dire news. As the pharmaceutical industry labored haltingly to get even a handful of new antibiotics to market, one was led to wonder: Weren't there *any* other hopes on the horizon?

There were, in fact, a few intriguing, if arcane, new approaches to drug-resistant bacteria—treatments that weren't drugs at all. They were, instead, natural substances from odd, unlikely places, being pitched by a cast of odd, unlikely characters.

13

HOPE IN FROGS AND DRAGONS

The first time he stalked a dragon, in November 1995, Terry Fredeking was scared. Bad enough to have had to fly all the way to Indonesia, deal with notoriously difficult Indonesian bureaucrats, brave the stifling heat, and find a local boat owner willing to whisk the biologist and two colleagues over to the sparsely inhabited island of Komodo. Worse, much worse, to lie in wait, awash with sweat, for the world's largest reptile to emerge from the forest in a hungry mood. That first time, Fredeking watched a Komodo dragon attack a goat. The Komodo was at least eight feet long and weighed well over 200 pounds. It looked like a dinosaur, Fredeking thought, it really did. It was all scales, with a huge mouth of large, curved teeth. One second it was lying in wait, all but invisible. The next, it was ripping out the terrified goat's stomach with a single bite. As it did, thick saliva dripped from the dragon's mouth, mixing with the blood and guts of the goat. *Ah, yes, the saliva,* thought Fredeking as he and his colleagues advanced from the bushes, tremulously holding long, forked sticks. The saliva was why they were there.

With luck, the dragon's viscous, revolting drool would contain a natural antibiotic that in some synthesized form could fight multidrug-resistant *S. aureus* and other bacterial pathogens. At the least, Terry

Fredeking, a genial, stocky, self-styled Indiana Jones from Hurst, Texas, would have the adventure of his life and possibly contribute to the fascinating new field of animal peptides. It sure beat collecting bat spit in Mexico and harvesting giant Amazonian leeches in French Guiana.

This latest approach to antibiotic discovery, tailor-made as it was for the Discovery Channel, traced in large part to a well-ordered lab at the National Institutes of Health. On a fragrant, early summer day in June 1986, a mild-mannered M.D. and research scientist named Michael Zasloff had noticed something decidedly odd about his African clawed frogs. As chief of human genetics at a branch of the NIH, Zasloff was studying the frogs' eggs to see what they could teach him about the flow of genetic information from the nucleus of a cell to the cytoplasm. He would inject genes into the eggs and see what happened. The frogs just happened to have large, good eggs for this purpose; their own biology was irrelevant to his work.

Some lab scientists killed the frogs after cutting them open to remove their eggs. Not Zasloff. He would stitch them up crudely—he was a pediatrician, not a surgeon—and, when enough of them accumulated in a murky tank in his lab, he would secretly take them to a nearby stream and let them go. On this particular day, Zasloff noticed that the tank appeared to have "something bad" in it, because several frogs had died overnight and were putrifying. But the frogs he'd just operated on, sutured, and thrown back in the tank appeared fine. Why was that? Certainly the frogs' stitches were not tight enough to prevent bacteria and other microbes from infiltrating their bloodstreams. Yet no infection occurred. No inflammation, either.

This was, as Zasloff put it later, his eureka moment, for even as he asked himself the question, he intuited the answer: the surviving frogs must have generated some substance that afforded them natural antibiotic protection. No likely suspects appeared under a microscope, so Zasloff began grinding samples of frog skin and isolating its elements. After two months, he still couldn't see what he was after. He could identify it, however, by its activity. He was dealing with two kinds of short amino acid chains called peptides—like proteins, but smaller. Scientists knew that peptides participated in many metabolic functions of living organisms, either as hormones or other compounds. They didn't know what Zasloff had just realized: that some peptides in frogs worked as antibiotics. Zasloff named them magainins—the Hebrew word for "shields"—and theorized that they might lead to a whole new class of human-use antibiotics. So promising was Zasloff's finding that when it was published a year later the *New York Times* devoted an editorial to it, comparing Zasloff to Alexander Fleming. "If only part of their laboratory promise is fulfilled," the *Times* opined of his peptides, "Dr. Zasloff will have produced a fine successor to penicillin."

Like Fleming, Zasloff had made his discovery through serendipity. It was a means about to become quaint. Soon genomics would begin to transform drug discovery into a high-speed, systematic search with state-of-the-art tools that analyzed bacterial DNA—the very antithesis of serendipity. But targeting individual genes, by definition, would yield narrow-spectrum drugs. No doctor wanted to rely exclusively on narrow-spectrum drugs, especially in the hours before a patient's culture was analyzed at the lab. Besides, a drug designed to hit one bacterial gene might soon provoke a target-changing mutation. Whole new kinds of *broad*-spectrum antibiotics were needed, too, and the best of those seemed less likely to be found by genomics than by eureka moments like Fleming's and Zasloff's, when a different approach presented itself as suddenly and clearly as a door opening into a new room. To date, virtually all antibiotics with any basis in nature had been found in soil bacteria or fungi. The prospect of human antibiotics from an animal substance suggested a very large room indeed.

The world had changed a lot since Fleming had published his observation about a *Penicillium* fungus, then basically forgot about it for more than a decade. Now biotech venture capitalists scanned the medical journals for finds that might be the next billion-dollar molecule. Zasloff would find himself swept from his NIH lab into the chairmanship of a new public company with Wall Street money and Wall Street expectations, his magainins hyped as the Next New Thing. Nearly $100 million later, he would also be the tragic hero of a cautionary tale about the challenges a maverick faced in bringing new antibiotics to market.

As an early postwar baby boomer at New York's Bronx High School of Science, Zasloff had thrilled to learn the periodic table, tinkered with home chemistry sets, and haunted the American Museum of Natural History. Magna cum laude at Columbia University, he had gone on to be mentored by Nobel Prize–winning molecular biologist Severo Ochoa at New York University Medical School, feeling equal pulls toward clinical medicine and academic science. One, as it turned out, informed the other. During his residency in the pediatric ward of Boston's Children's Hospital in the mid-1960s, Zasloff observed babies with cystic fibrosis, their airways phlegmy with *S. aureus, Pseudomonas aeruginosa,* and other bacterial pathogens. Tragically, those infections would soon destroy the babies' bronchi and lungs. The cause of cystic fibrosis remained a mystery, however, because the babies suffering from it had no apparent immunological deficiencies. Was it possible, Zasloff wondered, that those babies were lacking an aspect of their immune systems that no one had identified yet?

Zasloff chose science over medicine, finally, joining the National Institutes of Health as a public health alternative to fighting in Vietnam. But the mystery of the cystic fibrosis babies stayed with him. In 1981, a Swedish researcher named Hans Boman made a finding that seemed to bear out Zasloff's hunch. Boman wondered how insects, which lacked anything resembling a mammalian immune system, warded off routine infections. He determined that the silk moth *Hyalophora cecropia* secretes antibiotic peptides when exposed to bacteria. What especially interested Zasloff was that these peptides killed bacteria but not animal cells. Perhaps human babies had similar peptides—antibiotic defenses meant to kill bacteria but not their own cells. Perhaps the CF babies were missing these subtle defenses.

In fact, Boman was not the first to look for natural antibiotics. In the 1960s, a researcher at New York's Rockefeller University named John Spitznagel found that rabbits and guinea pigs had proteins that appeared to kill bacteria, but he was unable to get grants to pursue the matter. A decade later, Robert Lehrer, a doctor and professor at the University of California in Los Angeles, isolated two molecules from rabbits that killed, as he put it, "just about everything," but he too failed to get grants to pursue his work. Lehrer called the molecules defensins, a name that stuck, and observed that they were composed of three pairs of amino acids called cysteines. That led Lehrer to search for, and find, human peptides in 1983. Peptides had been overlooked, he realized, because they had no antibiotic properties when isolated. Put in with bacteria that were growing, however, they interacted with the bacteria in a way that killed the bugs. Lehrer saw the potential for new drugs in peptides but also wondered if giving a human patient more of the peptides already in the body would make much of a difference. "It seemed like carrying coals to Newcastle," Lehrer mused. Also, he balked at the challenges of going from academia into the drug business. Let someone else try it, he reasoned. So someone else did.

As he monitored their action, Zasloff discovered that magainins act not by targeting a bacterial protein, as nearly all modern antibiotics do, but by punching their way through the bacterial cell's membrane and forming ion channels that let water and other substances flow in. These, in turn, lyse the bacterium. This lysing occurred because magainins were positively charged and the bacteria had negatively charged elements called phospholipids on their membrane walls. The positively charged peptides homed in on the negatively charged cell membrane as if piercing an armored shell.

The wall-punching mechanism suggested that peptides might be especially useful against *resistant* bacteria. The proteins targeted by nearly all existing antibiotics could be changed or replaced, as Patrice Courvalin was just coming to see in Paris at the Pasteur Institute with vancomycin-resistant enterococci. For a bacterium to change its whole membrane would be orders of magnitude more difficult. It seemed impossible. And as far as Zasloff could see, peptides were drawn *only* to bacterial cell walls—never, in vitro at least, to the membranes of normal human cells. Which made them a perfect antibiotic.

Another NIH scientist might have published his findings, as Zasloff did, and gone back to tinkering in his lab with the next intellectual challenge. But as a pediatrician, remembering those babies with cystic fibrosis, Zasloff wanted to see peptides turned into drugs *right away.* His first step was to call the FDA. "I'm from the NIH and I just made a discovery that's about to be published," he told the bureaucrat he reached. "Can I get someone from the FDA to help me do what I have to do to make this into a drug?" The FDA had no system, it turned out, to help government researchers develop drugs while keeping their government jobs. Nor did the NIH have any such guidelines. (Not long after, the agency would allow researchers to profit in modest ways from technology transfer, but the burgeoning biotech industry would be filled with NIH refugees wanting a larger share of the proceeds of their discoveries.) Zasloff risked being fired or sued, he discovered, simply for fielding the calls that began to pour in after his article was published. If he talked to Merck, he could be sued by Bristol-Myers, because he was a government official obligated to favor no company over another. Finally, he got a call from venture capitalist Wally Steinberg. Steinberg was about to finance scientist Craig Venter's departure from NIH to start The Institute for Genomic Research.

Venter had insisted, as a condition of his own jump, that Steinberg make The Institute of Genomic Research a nonprofit foundation. Zasloff had terms, too. In addition to helping with the start-up—to be called Magainin—he wanted to teach, and he wanted to practice medicine as a pediatrician. The University of Pennsylvania had been wooing him for years. In short order, Zasloff became a professor of genetics and pediatrics there, in an endowed chair, as well as chief of human genetics at Philadelphia's Children's Hospital. For Magainin, set up outside Philadelphia in a corporate park of former farm town Plymouth Meeting, he worked as a part-time consultant.

It should have been an ideal setup, a dream life guaranteed to make any medical researcher sick with envy. But although Zasloff had thought he could work on peptides in his hospital lab and pass the results on to Magainin, the hospital's directors thought not. Work funded by the hospital, they declared, ought to remain the hospital's intellectual property. When the university, the third leg of Zasloff's new career, began lobbying for its own share of the proceeds, Zasloff gave up. Heartsick, he resigned a directorship at the hospital and gave back the endowed chair to the university. As of 1992, he would gamble his entire career on Magainin.

Since peptides seemed to work against almost anything, Zasloff and his colleagues scanned the market for a condition treated by only one drug: less competition, more opportunity. They settled on impetigo, the mild skin infection characterized by rashlike lesions and caused by skin bacteria, usually certain streptococci or *S. aureus*. SmithKline Beecham sold a drug for it called Bactroban, so Magainin could set up clinical trials identical to those SKB had performed for Bactroban—the FDA could hardly fail to approve those—and use Bactroban as the comparator drug. If the peptides worked as well or better than Bactroban, they'd be approved. From there, Magainin could go on to test peptides against more serious topical infections, have a couple profit-making products on the market, and so gird up for serious bloodstream infections.

The peptides sailed through phase one trials: applied to healthy human skin, they caused no harm. In phase two, they seemed to produce good results on forty-five people who actually had impetigo. The Bactroban trials had involved a placebo: simple soap and water. Magainin followed suit. But when the results of phase three trials were compiled in mid-1993, Zasloff was stunned. Though the peptides had done as well as Bactroban, neither product had done as well in his tests as soap and water! How, then, had Bactroban won approval in the first place? Zasloff never learned. The FDA merely announced that peptides had failed the trial. Overnight, Magainin's stock plunged from eighteen to three dollars a share.

As Magainin teetered on the verge of collapse, Zasloff desperately pulled a rabbit out of his hat. Or, rather, a dogfish shark.

By 1993, inspired by Zasloff's original paper, dozens of other scientists had gone in search of peptides in other animals. They'd found them just about everywhere they'd looked—seventy different antibiotic peptides in all—in everything from insects to cows to Komodo dragons.[1] Intriguingly, different creatures secreted peptides from different kinds of cells. Many insects made them in their white blood cells. In horseshoe crabs, they appeared in the blood elements called platelets. In the frog, as Zasloff had determined, they appeared in a part of the nervous system called the granular glands: the frog empties these glands, Zasloff found, when it is stressed or when the skin is torn. As for humans, they turned out to harbor peptides of their own, just as Lehrer and Boman had observed: in white blood cells, in the gut, and, notably for cystic fibrosis babies, in certain cells of the airway called the ciliated epithelium. Perhaps, thought Zasloff, some other animal's peptides would make a more potent antibiotic than those of the African clawed frog—potent enough to bring investors scurrying back to Magainin.

One day Zasloff gave his standard stump talk about peptides to a group of scientists at the Marine Biological Laboratory in Mt. Desert, Maine. John Forrest, a professor at Yale University's medical school, raised his hand to say that he'd spent nineteen summers studying the dogfish shark, and, by God, if the African clawed frog had peptides, so must the shark. The shark had long been Forrest's experimental animal model, as the frog was Zasloff's. Small and hardy, the shark had large, simple cells and organs that made it easy to study. Best of all, when Forrest operated on a dogfish shark, he could suture it up and toss it back into a tank of dirty water, as Zasloff did with his frogs. Inevitably, the shark healed without infection. Zasloff went home with a shark stomach to study, expecting to find peptides. Instead, he found a new kind of steroid with even stronger antibacterial action—yet another element of the innate immune system. He called it squalamine. "Hey!" he told Forrest by phone. "Send me more of those shark stomachs!"

---

[1] Eventually, hundreds more would be identified. Robert E. W. Hancock of the University of British Columbia in Vancouver, Canada, would publish, on his own, some 300 papers on different peptides.

Eventually, Zasloff found a way to purify shark squalamine and switched to livers, because a company in New Hampshire could FedEx him half a ton of them a week. Zasloff himself would wheel the heavy boxes of stinking shark organs in from the loading dock, then start slinging them into a giant meat grinder. The purification process involved heating the ground livers in garbage cans like great vats of soup, skimming the squalamine-rich scum from the top, then filtering the scum through a more high-tech set of steps.

Along with squalamines, Zasloff found other steroids in the purified gunk. He figured there were more than twelve kinds in all. Each had broad antibiotic effects, but each also seemed to target a specific kind of cell in the shark's body. Publication of the finding had brought calls from around the world, and these helped focus Zasloff's study. Several of the steroids, he found, worked as anticancer agents both in dogfish sharks and in humans. One kind even prevented lymphocytes from carrying out the AIDS virus's orders to make more virus.

Certain that he had found a way to save his company, Zasloff contacted Anthony Fauci, Director of the National Institute of Allergy and Infectious Diseases at the NIH and, as such, the top U.S. government official involved in fighting AIDS. Fauci established a Cooperative Research and Development Agreement, or CRADA, with Magainin, and Zasloff started injecting squalamines into AIDS-infected mice and dogs and monkeys. The squalamines worked brilliantly—up to a point. They stopped the growth of the lymphocytes, just as they had in vitro. Unfortunately, as soon as the animals were hit with the squalamines, they stopped eating and began to lose weight.

For months, Zasloff struggled to solve the dilemma. A lonely figure reeking of shark liver, he spent his days skimming scum and injecting steroids into AIDS-infected lab animals. No approach worked. The animals' lymphocytes stopped growing, as did the AIDS virus, but the animals simply would not eat. In Washington, Anthony Fauci gave up hope: the prospect of halting a patient's AIDS infection only to induce death by starvation was obviously unacceptable.

Okay, Zasloff declared at last, *okay.* All was not lost. "What nature has given us," he announced to his devastated colleagues, "is an *appetite suppressant.*"

Zasloff had two strikes against him, and, as far as his backers were concerned, it was the bottom of the ninth. But by the mid-1990s, the sharp rise in resistance around the globe had cast peptides, his other finding, in a more favorable light. Peptides still appeared utterly impervious to all the new mechanisms of resistance that bacteria had employed. Intrigued, the FDA offered to let Magainin try peptides once more, this time on a more serious topical condition than impetigo: infected diabetic ulcers. As the FDA knew, the existing antibiotics used against these painful foot lesions caused such debilitating side effects that patients usually stopped taking them—though the lesions, when infected, tended to invade muscle and bone and even led to amputation of the affected limb. Now, in addition, resistance to these antibiotics was rising. Worse, the most promising of them, Trovan, would soon be pulled from the market for causing liver toxicity. Here was a real need—and market niche—that peptides seemed perfect to fill.

Because patients could suffer irreversible harm from diabetic ulcers, the FDA ruled that no placebo would be needed. Zasloff's peptides merely had to do as well or better than one of the comparators, a fluoroquinolone called ofloxacin, which came not as a topical ointment but in oral form. Magainin breezed through phase one trials: the peptides, as shown in the previous trials, caused no harm to the skin of healthy people. To speed the process, the FDA let Magainin combine the next two phases. Roughly 1,000 patients were recruited from more than fifty medical centers in the United States between 1995 and 1997. These were very sick patients, their lesions excruciatingly painful. When doctors swabbed the lesions with a peptide solution, most of the patients seemed to improve.

As Zasloff pored over the final results, he felt encouraged, if not wildly optimistic. The topical peptides had not quite outperformed oral ofloxacin, but they'd done nearly as well. Certainly the tests had shown that MSI-78, as Magainin's latest peptide was known, had a broad and powerful spectrum, did not provoke resistance, and had no direct side effects. The results were strong enough for SmithKline Beecham to sign on as a partner. SKB would market the product as Locilex. Now all Magainin needed was formal approval by an FDA advisory panel.

The panel, composed of seven experts from various fields, met on March 4, 1999, in Silver Springs, Maryland, to spend the whole day debating the merits of Locilex. Zasloff, looking on from the audience of 300, thought the morning session went well, though he was a bit disconcerted by the early testimony of an FDA-appointed expert brought in to give the panel a background lesson on diabetic ulcers. To Zasloff's dismay, the expert spoke only about uninfected diabetic ulcers, which were the sort the expert had treated. In his opinion, *un-*infected ulcers needed *no* antibiotics. He seemed to leave the panel with the impression that only the most advanced cases of infected ulcers would need treatment by antibiotics and that in those cases a topical agent would hardly suffice. By the lunch break, however, a long, careful presentation of Locilex's strengths by two of Magainin's executives seemed to have set the balance right.

Perhaps the panel members were served an inedible lunch. Perhaps the meeting room was too hot or cold. Whatever the reason, the members reconvened in a grumpy mood. One of the seven declared that in her opinion—founded not on clinical experience, only on the morning's thirty-minute tutorial—no antibiotics were needed for infected diabetic ulcers. "Just cut the infected tissue out and throw it in the garbage can," she declared. One after another of the members agreed. The panel's chairman, Dr. William Craig, pointedly disagreed, as did Dr. Barbara Murray, the University of Texas's expert on enterococcus whose letter to the European Union endorsing the concept of animal-to-human transfer of VRE resistance had caused such jitters. Nevertheless, the final vote was seven to four, a decision upheld formally by the FDA some months later. For lack of two sympathetic minds, Michael Zasloff's thirteen-year crusade to use peptides against drug-resistant bacteria was finished.

There were those, in the wake of Magainin's defeat, who muttered

that peptides never would have worked for systemic infections anyway. "You could show in the test tube that a peptide of the sort
Zasloff was using would punch a hole in the membrane," one rival declared. "But it would also punch a hole in *any* membrane. It's nondis-
criminating. It would rupture red blood cells and cause hemolytic
anemia. So there's been much ado about nothing." Zasloff denied
that. "Those peptides are extraordinarily specific in vitro," he declared. "And if one unit of peptide kills a bacterium, we found it takes
1,000 units to damage a human cell." But in vivo, he admitted, that
changed. An awful lot of frog peptide, it turned out, was needed to
kill a laboratory animal's bacterial infection, almost as much as would
make the peptides toxic. Apparently the lab animals had other defenses that perceived frog peptides as foreign agents and so targeted
them. "Frog peptides worked for frogs," observed one medical entrepreneur. "But we're not frogs, are we? And we're not sharks."

Over the next two years, Zasloff himself came to wonder if animal
peptides would ever work in people. Perhaps the way to go was to focus on *human* peptides—now that plenty of those had been found—
and to try to strengthen the barrier of innate immunity to fight human
infections.

In a desperate bid to keep his company alive, Zasloff pushed
squalamine into clinical trials as an appetite suppressant. He was serious. It was the Hail Mary play, as he put it, that might save the day.
But no one else seemed to believe he could pull it off. Meanwhile,
SmithKline Beecham wrote off its partnership on Locilex.

In the fall of 2000, Zasloff's own directors lost faith. The scientist
whose discovery had inspired the company was made a consultant—
pushed out, as Zasloff later admitted—and the corporate direction
changed. The clinical tests with squalamine as an appetite suppressant were carried on: the stuff did look promising, wacky as the route
to its application may have been. Early results had shown squalamine
to be effective, as well, against ovarian and non-small-cell lung cancer.
But in corporate press releases, no further mention was made of
antibiotics—or peptides. From now on, the company would use genomics to find new targets and new natural substances like hormones

as drugs. To make that perfectly clear, the name Magainin was changed to Genaera.

Zasloff felt hurt, and he felt angry, but he was still determined to see peptide-based antibiotics reach the market. In late 2001, he agreed to head up a new company based on work by a group from the University of Leiden in the Netherlands. The group was using a set of human peptides which seemed, in vitro, to have very weak antibiotic activity. Yet in vivo they became incredibly potent. This was such a counterintuitive approach, as Zasloff put it, that he'd never thought to pursue it. These molecules, he reported excitedly, were perfectly tuned for their role in the human body. They distinguished bacterial cells from human cells and did no damage to the latter. Not only that: they actively sought out the site of infection, like heat-seeking missiles. By the same approach, Zasloff hoped to boost animal peptides as animal antibiotics. Imagine, he said, how well they would serve as . . . *growth promoters*.

In his more contemplative moments, Zasloff admitted he'd made some grand mistakes. But he had no regrets about his role in establishing a burgeoning new field: some 3,000 articles on peptides had been written since his seminal paper of 1987, some 500 peptides discovered. The innate immune system was now part of science. And for Zasloff, the most promising aspect of peptides was still their potency against resistant bacteria. They'd persisted through most if not all of evolutionary history. In all that time, bacteria had never become resistant to them. Was it too much to suggest that they constituted the Achilles heel of pathogens? That bacteria never *would* become resistant to peptides? "They've had a billion years to fend these things off," Zasloff said, "and this is what we've got."

As the president of Antibody Systems, a small, Texas-based biotech company, Terry Fredeking had dedicated himself to the search for peptides and other natural substances in animals, the more exotic the better, that might lead to drugs for resistant pathogens. Michael Zasloff's discovery had made his work possible; one of Zasloff's former students was in his employ. Some of his samples—which included parasites from Tasmanian devils, among other odd things—showed promise in vitro, but Fredeking hungered for more. In truth, he was a bit of a showboater, eager to make his name, with the sort of chutzpah that made lab scientists shudder but sometimes got things done. "There's got to be something bigger than this," Fredeking said one day to one of his consultants, Professor George Stewart of the University of Texas. "What can we do next that's dangerous, exciting, and will advance science?"

"How about Komodo dragons?" Stewart suggested.

"Komodo dragons?" Fredeking echoed. "What in the heck are they?"

Stewart explained that the world's largest lizard, formally known as *Varanus komodoensis,* was justly famous for being one of a handful of predators big and fearless enough to prey on human beings on a somewhat regular basis. In fact, humans were by no means its largest prey: full-grown Komodos were known to bring down 2,000-pound water buffalo. Found only on the Indonesian islands of Komodo, Flores, and Rinca, the dragons were descendants of mososaurs, massive aquatic reptiles that roamed the seas 100 million years ago. Though the Komodo dragon did often hunt down and devour its prey, it also had a craftier method of killing that hinted at the presence of antibiotic peptides. A stealth hunter, the dragon lay in wait for sambar deer, crab-eating macaque monkeys, and other mammals of its habitat, then lunged for the abdomen of its passing prey with toothy jaws as strong as a crocodile's. Almost always, its wounded victims escaped, because the dragons, many of them heavier than a fat, six-foot-tall man, could run only in short bursts. But because the dragons often feasted on rotting carcasses, their jaws teemed with virulent bacteria. Within seventy-two hours of being bitten by the great lizard, animals would die of bloodstream infections brought on by these bacteria. Eventually the dragon would come lumbering over to take his meal at last.

Both because of its lethal saliva and because the dragon ate carrion teeming with *more* bacteria, zoologists had long wondered what made the dragon immune to all these pathogens. Whatever it was had to be really powerful, because of an evolutionary oddity about the dragon's teeth. Razor sharp as they were, and serrated like a shark's, the dragon's teeth were actually covered by its gums. When it snapped its jaws shut on its prey, the teeth cut through the gums. The dragon's lethal saliva, then, had access to its bloodstream. Yet the Komodo remained uninfected.

"In all likelihood," Stewart finished, "the dragon's bacteria has been battling with its immune system for millions of years, with both sides getting stronger and stronger over time to keep each other in balance."

"That's it!" Fredeking exclaimed. "Lead me to 'em!"

Nearly three years passed before Fredeking and two colleagues could secure permits to take samples of Komodo dragon saliva. Both the Indonesian and the U.S. governments had to be petitioned, because the dragon is an endangered species, and most of the 6,000 that remain are found within Komodo National Park, which covers several islands and is now a World Heritage Site. Finally, on November 30, 1995, came the momentous day. Fredeking and Jon Arnett, Curator of Reptiles at the Cincinnati Zoo, flew to Bali, where they met up with Dr. Putra Sastruwan, a biology professor and Komodo dragon specialist at Bali's University of Udayiana. They took two days to recover from jet lag, then flew to the Indonesian island of Flores in a small Fokker plane that made Fredeking more nervous than the prospect of facing Komodo dragons.

The next day they crossed over to Komodo by ferry—another unnerving experience for Fredeking, since the ferry had sunk on several occasions. From a distance, the island appeared shrouded in fog, with protruding volcanic cliffs. Close up, Fredeking saw that its coastline was lined with rocky headlands and sandy bays. Much of its interior was dry, rolling savanna, with bamboo forests halfway up the larger peaks. The island supported a variety of large mammals, all imported by man: deer, water buffalo, boar, macaque monkey, and wild horse. No one knew how the Komodo dragons had come to the island. Paleontologists believed their genus evolved in Asia 25 to 50 million years ago—as reptiles, but not dinosaurs—then migrated to Australia when those two land masses collided. Because Indonesia lay closer to Australia at that time, the dragons may have swum to the islands and proliferated, growing larger over time, because the islands contained no predators for them.

Hot and sweaty, the biologists spent their first night on the island in a village that was nothing more than a cluster of bamboo huts. Over a local dinner of rice and fish, they heard stories of the dragons' ferocity. Eight villagers, mostly children, had been attacked and killed by Komodos in the fifteen years since the national park was established and records began to be kept. One old man had paused beside a trail to take a nap: his supine form looked vulnerable and inviting, and he, too, fell victim to a dragon's steel-trap jaws. Other stories, unverifiable, had circulated ever since W. Douglas Burden had come over in 1926 on behalf of the American Museum of Natural History and made a first formal study of the beasts, capturing twenty-seven of them and naming them Komodo dragons. Burden also brought the first Komodo dragon back to New York City. He told the story of his adventure to Meriam C. Cooper, among many others, and fired the Hollywood producer's imagination. Cooper changed the dragon to an ape, added Fay Wray, and in 1933 gave the world *King Kong*.

It was the next morning that Fredeking saw a Komodo dragon rip open the belly of a terrified goat. He had briefly considered bringing tranquilizer guns to bag his prey, but scotched the idea when he learned that a sedated dragon is likely to be eaten by his peers. Komodos are cannibalistic; they will eat each other, including their own young. Newly hatched dragons know, by biological imperative, to scamper immediately up tall trees and spend their first two years as arboreal creatures, safe from the snapping jaws of their parents below.

Instead of using sedatives, Fredeking and his cohorts emerged from their hiding places with long, forked sticks and one long pole designed for catching crocodiles: an extendable pole with a wide noose at the end. The noose was slipped over the dragon's head and pulled tight. Before the befuddled creature could react, six men jumped on him. Jon Arnett held the dragon's head and began wrapping duct tape around it. Others wrapped tape around its extended claws. Equally important, a ranger grabbed the dragon's powerful tail. Fredeking reached for the long swabs he'd brought for collecting the dragon's saliva. He looked at the dragon's furious eyes and then, startled, at its *third* eye: a "parietal" eye in the roof of its cranium, which acts as a light-sensing organ. He dabbed at the saliva, shocked at how thick and viscous it was—like petroleum jelly. One sample was slipped into a vial, then another. Fredeking began to feel euphoric. That was when he heard one of the others say, in real terror, *"Oh my God."*

Fredeking looked up, and felt the paralyzing fear of the hunter who has gone from being predator to prey. More than a dozen Komodo dragons were advancing from all sides. Drawn by the noisy struggle of the dragon that had been captured, the lizards had converged with the quaintly Komodian hope of eating it—along with the men around it. Panting with adrenaline, the men pushed at the dragons with their forked sticks. With their length, body mass, and sheer reptilian power, the dragons easily could have pushed right up to the men and started chomping away, either at the duct-taped dragon or at the hors d'oeuvre plate of tasty human legs. But the sight of tall men with sticks seemed to confuse them. One of the park guards—an old hand at dealing with the dragons—aggressively advanced on one of the larger lizards and pushed him away with his forked stick. For a tense minute or so, the outcome remained uncertain. Then, one by one, the dragons turned and clumped away, including the first creature from which the sample had been taken. Fredeking took a long breath. "Man, oh man," he said. "What we do for science."

On that first trip, both of Fredeking's cohorts incurred deep scratches on the insides of their calves by sitting on the dragon's back to help restrain him. They knew that the dragon's scaly skin—as scaly as chain mail—was rife with bacteria, too. Within hours, they were infected and running fevers. Fredeking was running a fever, too. All three took ciprofloxacin and soon felt better. Not surprisingly, the dragon's bacteria were susceptible, given that the bugs had probably never encountered fluoroquinolones or any other commercial antibiotics.

Along with saliva swabs, Fredeking came away with samples of blood from the dragon's bleeding gums. Flash frozen in liquid nitrogen and stored in Thermos-like containers, the samples were flown back to Texas, where Fredeking's researchers got to work. They counted sixty-two different kinds of bacteria in Komodo saliva. The most potent of the lot was *Pasteurella multicida,* common in many domestic animals, though in far less virulent strains. They found antibiotic peptides, too, along with a small molecule that did an even better job of killing bacteria. In vitro, the molecule knocked out MRSA, VRE, and *E. coli* 0157:H7. Dr. Don Gillespie, a veterinarian in touch with Fredeking because of his work with Komodos at the Nashville, Tennessee, zoo, worried that the peptides might not last long in the human body. Still, if this new small molecule went unrecognized by human antibodies even for a short time, it could be a perfect candidate for a new class of antibiotic.

First, the researchers would have to try the peptides, and the molecules, in mice, then guinea pigs, then primates. And even the gung-ho Fredeking knew better than to make any predictions. "If it makes mice grow long green tails and crave human flesh, we'll know it's not good," he said. "Basically, anywhere along the trail here, this thing could fall apart."

As stock tips, peptide companies probably ranked just above the average dot.com. More promising—perhaps much more promising—was another natural substance, the antibiotic properties of which had first been observed nearly a century before.

14

BACTERIA BUSTERS

On a frigid day in January 2001, a forty-five-year-old French Canadian jazz musician named Alfred Gertler settled, with some effort, into his seat on a British Airways jet bound from Toronto to London, with a connection to the former Soviet Republic of Georgia. He winced with obvious discomfort and spent a minute adjusting his bandaged foot. When a flight attendant asked if he was all right, he assured her he was fine. He chose not to tell her that he had a deep, suppurating, resistant staph infection on his bandaged ankle and that he was traveling more than halfway around the world in a desperate effort to eradicate it.

Customs in Tbilisi, Georgia, as expected, was a dreary process in a sparsely furnished terminal building dark without daytime electricity. The wide avenues leading in from the airport were just as grim, lined on either side by featureless cement apartment blocks from the Soviet era. Some of Tbilisi's best buildings were pockmarked with bullets or devastated by mortar shells, the aftermath of Georgia's civil war of the early 1990s and stark testimony to the now independent country's ongoing financial straits.

From a distance, the Eliava Institute of Bacteriophage, Microbiology, and Virology appeared stolid and unaffected by the turmoil of the last decade. But as Gertler approached the institute's wooden doors, he could see that they were so warped they barely shut. Inside, the once grand foyer was dingy with peeling maroon paint and, like the airport terminal, dark without electricity. A glass exhibit case stood cracked and empty. After a moment or two, Gertler realized that the building was also without heat.

Gertler's staph saga had begun well over a year before, when he signed on as bass player for a house band on a cruise ship headed to Central America. During a stopover in Caldera, Costa Rica, Gertler had gone mountain climbing and sustained a bad fall. His ankle swelled immediately; at a local hospital, x-rays confirmed he had a compound fracture. The ship had to proceed without a bass player while Gertler remained in the hospital to have his fractured bones set in surgery with compression screws, and a cast put on. By the time he got home to his wife, the ankle was infected, and Gertler was feeling the onset of staph infection symptoms: fever, fluid, and swelling from the infected area, and general exhaustion. The infection, as he put it, "just really climbed over" him. Had he contracted the staph bacteria from the Costa Rican hospital? Or had it resided on his skin until his injury gave it access to his bloodstream? Gertler would never know.

Back in Canada, the musician was treated over the next six months with ciprofloxacin and another antibiotic through an intravenous line directly into his bloodstream. Neither stanched the infection or did much to ease the symptoms. Gertler felt like a walking corpse. His doctors were convinced that his staph infection was only mildly resistant, yet it remained all but inaccessible to systemic antibiotics. For nearly another year, Gertler was put on oral antibiotics. Sometimes the infection would subside a bit, but when Gertler started working, keeping a jazz musician's late hours, it would flare up again. It was actually two infections, one on either side of his ankle, with a golf-ball-sized cavity of infection in between, on his ankle bone. When Gertler learned about vancomycin, he asked to be treated with it, but his doctors demurred: they were saving it, they said, for more severe and resistant infections. As for Synercid and Zyvox, Gertler had never heard of them and to his recollection his doctors made no mention of them.

One day, Gertler went to visit a friend whose own more severe staph infection had landed him in the hospital. At his friend's bedside was a newspaper article about something called bacteriophages. Gertler was fascinated. He started calling around. Eventually, through a Web site on phages maintained at Evergreen State College in Olympia, Washington, Gertler learned about the Eliava Institute, a seventy-year-old center of bacteriophage research. He asked his doctors what they thought of his going over to Tbilisi to give phages a try. They shrugged. What did he have to lose?

In Tbilisi's Republic Hospital, Dr. Guram Gvasalia began treating Gertler in a fairly holistic way, with a number of therapies: enzymes, an acupuncturelike laser therapy, some kind of electrical charge treatment. Then Gertler began receiving a phage product called PhageBioderm, which was inserted deep into the wound. Another kind of phage was used topically for hours at a time. The treatment wasn't painful; Gertler just felt cold, because the hospital appeared to be minimally heated and to have water just one or two hours a day. As he lay wide-eyed and shivering in his bed, Gertler heard the sound of occasional gunfire resounding through the city. As a jazz musician, he'd led a pretty unconventional life. But this was definitely the weirdest experience he'd known.

At the same time, it was intensely gratifying. The doctors and nurses treated Gertler with great tenderness and concern, shrugging off their own bleak circumstances. (The average salary for doctors at the hospital, Gertler learned, was about thirty dollars a month.) They administered treatments that Gertler calculated to cost upwards of $10,000, yet only charged him $300—and apologized for that. Best of all, after two days, Gertler's ankle infection stopped draining. After a week, there was no evidence of the staph infection at all. It was *gone*.

Alfred Gertler may well have been the first westerner to fly to Tbilisi for bacteriophages to cure an infection. But in the decade of growing multidrug resistance, with so few antibiotics in development, a number of doctors and scientists—Glenn Morris among them—had begun to wonder if a very different kind of bacteria fighter, long discredited, might be worth another look. They, too, had begun to make pilgrimages to Tbilisi, to follow Georgian researchers, bundled in winter coats, down dank, high-ceilinged halls to laboratories with equipment decades old, to study plates of bacteria and phage kept in agar that the researchers tipped out from old beer and vodka bottles, and to hear of astounding results when a certain phage was matched with just the right strain and species of bacteria.

Most scientists still rolled their eyes at the very mention of phages. But their existence, at least, was beyond dispute. Phages are viruses. At one fortieth the size of most bacteria, they may be the most numerous organisms on the planet. Their raison d'être is to attack bacterial cells—in soil, in water, in the guts of animals and man—in order to proliferate. Viewed through an electron microscope, phages look like live lunar landers, with multipaneled heads, spindly legs, and tails they use as natural syringes. When a phage lands on a bacterial cell, its legs attach to its surface. Through its tail, it then injects DNA from its "head" directly into the cell. What happens next depends on whether the phage is temperate or lytic.

Temperate phages colonize the cell and hang out. Later, if the cell is threatened—by an antibiotic or by any worrisome change in its environment—the temperate phage may decide to bust out. Then its DNA commandeers the bacterial cell's genetic machinery, forcing it to make "daughter" phages instead of more bacteria. The daughters burst through the cell, which is to say they lyse it, so that the cell is destroyed. And then on the daughters go to find another, more hospitable cell.[1]

---

[1] Geneticists often put genes of their own choosing into temperate phages and have the phages inject *those* into the phages' normal host bacterium. The bacterium then takes up those substitute genes and, when induced by an environmental change, starts replicating them. Geneticists can then "manufacture" as many genes as they need. This is the basis of genetic engineering.

Lytic phages, by contrast, burst the bacterial cell almost immediately. As soon as the cell begins to replicate, they take over its genetic machinery. Within the bacteria's short replication period, as many as 200 daughter phages are made, their own heads and tails clamping together like so many Snap-It tools. These daughters of lytic phages lyse the cell and go off to alight on new bacterial cells, injecting them with more phage DNA and causing them to lyse, too.

Scientists now know that in a marvelous exercise of ecological balance, virtually every strain and species of bacteria appears to be dogged by lytic bacteriophages that have evolved specifically and precisely to attack *it*. For example, a lytic *S. aureus* phage will do a fast and brutal number on *S. aureus* but have no effect whatsoever on *E. faecium*. For Felix d'Herelle, the seeming unpredictability of phages was a maddening frustration. Still, in 1917, when he coined the word *bacteriophage* to define these tiny bacteria "eaters," he had absolutely no doubt that they were, in fact, viral predators that might cure bacterial infections. His confidence was especially striking because, in this time before electron microscopes, he couldn't *see* the organisms he was christening. He just knew they were there.

Though less celebrated than Louis Pasteur in his lifetime, and all but forgotten half a century after his death, Felix d'Herelle is in some ways a more dramatic character, a true tragic hero whose obsessiveness led him to a brilliant insight, only to get the better of him and leave him, at the end of his life, a pitiable and broken man. Others contributed to the discrediting of phages as antimicrobial agents, but d'Herelle did the lion's share of damage himself, with incorrect claims that would have made Pasteur shudder. His reputation as a crank lingers even now, complicating the various efforts to get bacteriophages into clinical trials.

In an old black-and-white photograph taken before his star began to fall, d'Herelle looks as one imagines Inspector Maigret might look, or a character out of *Tintin:* an intense fellow peering through wire-rimmed glasses, with a dark twirled mustache and a beard trimmed to a point. Born in 1873 in Montreal, he moved to Paris at six years old, after his father's death, and was enrolled by his Dutch-born mother in a proper lycée. At seventeen, however, his schooling ended, and with a generous gift of 3,000 francs from his mother he began a life of travel that soon engaged him in scientific research and work in far-flung places—places, suggested later critics, where his lack of a college degree would be no impediment.

It was on his first trip, to South America while still a teenager, that d'Herelle resolved to become a microbe hunter. After exploring the Canary Islands and hunting small game with a pistol in Paraguay, he boarded a ship in Rio de Janeiro for the trip home to Paris. Barely had the *Royal Mail* left port when a yellow fever epidemic broke out on board, eventually killing twenty. The young d'Herelle listened to a ship's doctor confess that medical science remained baffled by every aspect of the disease, its cause, its route of transmission, its cure. D'Herelle resolved then and there to join in the war against infectious diseases. "It is probable that I have, by birth, the first required quality needed to make a great microbe hunter," he wrote later with characteristic immodesty of his experience amid the dying on that trip. "Most of the passengers were in anguish: I was perfectly calm, I thought I was invincible."

After marrying at twenty and settling in Montreal for the birth, the next year, of his first child, d'Herelle set up a home laboratory and began studying microbes with the help of a professor at the local university. Barred, without a degree, from getting any formal job in science or medicine in his native Canada, he relied instead on the good graces of a family friend high placed in the Canadian colonial government. The friend proposed hiring d'Herelle to determine how to ferment and distill Canadian maple syrup into whiskey for sale in the United States. "Pasteur made a good beginning by studying fermentations, so it might be interesting for you, too," the friend said. D'Herelle made some progress, but before he could go further, the market changed. From fermenting he went to chocolate-making, but his efforts ended in bankruptcy. Desperate, d'Herelle responded to a magazine advertisement by the government of Guatemala for a microbiologist. Whatever he claimed as his education, he got the job, the first of several in far-flung places that would lead him to bacteriophages.

Years later, when rivals accused him of backdating his findings about phages in order to claim credit as their discoverer, d'Herelle would point to early work done in these exotic locales, specifically with locusts in 1910 in Merida, Mexico. In trying to help Mexico's government devise some tool to stop an invasion of locusts, he would recall, he had studied the black, pastelike diarrhea of sick locusts to see what organism might be killing them. The locusts were afflicted, he found, by a bacterium that caused lethal inflammation of the intestine, or enteritis. The bacteria were harmless to humans, he found, so they might be used in a then novel way: as a means of biological pest control. In some cultures of the bacteria, however, there were clear spots, indicating that something—some organism too small to be seen under a microscope—was eradicating the bacteria.

To learn more about his bacteria, or coccobacilli as he called them, d'Herelle moved his family to Paris and took an unsalaried position at the Pasteur Institute—the lack of a salary suggesting that, to the Pasteur's directors, he probably admitted his lack of formal education. Still, his work interested director Emile Roux enough to get him sent to various postings for the institute over the next several years—Argentina, Algiers, Turkey, and Tunisia—where d'Herelle used his coccobacilli in one locust infestation after another. He claimed great success for the approach, though one Argentine commission tested his coccobacilli and declared the results "completely negative," and a colleague at the Pasteur Institute's outpost in Tunisia declared his efforts there a failure. Perhaps, d'Herelle thought, the coccobacilli failed because those invisible organisms he'd first noted in Mexico killed the bacteria before the bacteria could kill the locusts.

The First World War was well under way when d'Herelle made the profound observation that these invisible organisms could kill bacteria in people, too. He was back at the Pasteur Institute, still as an unpaid volunteer, preparing bacterial vaccines for the Allied troops and

doing microbiological analysis of dysentery isolates from sick soldiers. Between July 20 and August 15, 1915, ten infantrymen and two
civilians were afflicted with a strain of hemorrhagic dysentery so severe that their attending doctor made a formal report of it. The particular strain of *Shigella* bacteria causing the dysentery was different
from any that d'Herelle had ever seen, so he stored fecal samples in
flasks and tested cultures from them on a continuing basis. One
morning, he came in to find that a flask that had been cloudy the night
before was perfectly clear. On the bottom was a little film of white bits
that turned out to be bacterial debris. Something in the flask had
killed the *Shigella*—something, he quickly determined, very much
like the agent that had killed the locusts' coccobacilli.

Whatever had killed the bacteria in culture, d'Herelle intuited,
might also have killed the bacteria in the intestines of the soldier
whose fecal sample was stored in that flask. D'Herelle sought out the
soldier. Sure enough, his fever had subsided, his symptoms abated.
Now d'Herelle went to the flask and poured it through a filter so fine
that no bacteria could get through it. When a drop of that residue
cleared another flask of cloudy *Shigella* bacteria, d'Herelle began to
get excited. When he filtered the second flask and watched the
residue clear a *third* cloudy flask, he knew he had proof that his microbe existed, and that it did kill bacteria. In a crisp, two-page paper
that startled the scientific world in 1917, d'Herelle wrote, "I have isolated an invisible microbe endowed with an antagonistic property
against the bacillus of *Shiga*." In the paper, he called his germ a bacteriophage, the name that stuck.

On the battlefields of World War I, mechanized warfare was mowing down millions of soldiers, yet even more millions were dying of
battlefield infections in this preantibiotic era. The prospect of a natural organism that might fight *Shigella* electrified the medical field. All
the more amazing was d'Herelle's confident speculation that bacteriophages might kill other bacteria, too. In fact, he soon theorized,
bacteriophages were the body's *entire* defense against pathogens.

Amid the press acclaim that made the forty-four-year-old d'Herelle an international figure and the thousands of medical papers on phages generated like so much ticker tape by his discovery, a number of influential colleagues began to carp. They started with his claim that phages constituted the body's immune system. Where, muttered Pastorians, did that leave Elie Metchnikoff and his brilliant theory of phagocytes, the white cells that destroyed toxins in the body? Or Emil von Behring and his use of antitoxins in the blood to fight diphtheria? At the same time, they scrutinized d'Herelle's claim that phages were viruses. How could he know this? Why couldn't the invisible microbe, as he called it, be an enzyme instead?

Beleaguered and defensive, d'Herelle left the institute to pursue his phage experiments in the French countryside, muttering that the world would come around soon enough. But even he hardly expected it to do so as soon as it did. In 1920, a Georgian microbe hunter as zealous as d'Herelle undertook a series of experiments at the Pasteur that validated phages as antibacterial therapy. With that, he became d'Herelle's intellectual soulmate in the search for phages, and his partner in the dream of establishing a phage center in his native Tbilisi.

George Eliava had discovered phages on his own, the same year d'Herelle published his two-page paper in Europe, by observing their activity on a sample of river water. The Mtkvari River originated in Turkey and flowed through several villages in both Turkey and Georgia before bisecting Tbilisi. As head of Tbilisi's Central Bacteriology Laboratory, Eliava wondered if the Mtkvari might be the source of the cholera that had swept through the region with depressing regularity. Eliava collected a specimen of the river water to see if it contained the bacteria that caused the dread disease. He placed a few drops of the water on a glass slide and carefully placed it under his microscope lens. His hunch was right: the slide contained an ample population of cholera bacteria. Just as he had begun to observe it in earnest, he was summoned to an urgent meeting. He left the slide in position under the microscope lens; by the time he got back to it, after three days of managerial duties, he assumed the slide had dried out. To his surprise, the glass slide cover he had placed over the specimen of river water had kept the water from evaporating. More surprising, the cholera bacteria were gone. They could hardly have escaped. What had happened? He repeated the experiment several times. Like d'Herelle, Eliava theorized that some tiny microbe, invisible through the microscope, had killed the bacteria. But not until he arrived at the Pasteur Institute in 1918 for an internship in microbiology and heard the eccentric d'Herelle's claims about bacteriophages did he realize what had happened.

Eliava was as brash and dramatic a character as d'Herelle, but a lot more charming. Unlike d'Herelle, for example, he had a sense of humor. Also unlike the intense Canadian, he adored the arts, good wine, and beautiful women. As gregarious as he was intelligent, the round-faced Eliava soon ingratiated himself among the cliquish Pastorians. Before long he felt respected enough to petition director Emile Roux for permission to repeat d'Herelle's controversial experiment. Permission was granted, and the experiments were successful. So impressed was Roux that he sent d'Herelle a telegram—d'Herelle was in the French countryside, feeling sorry for himself—informing him of the good news. The French Canadian rushed back to Paris. Entering the doors of the institute, he began to shout: "Where is this Georgian! Show him to me!" When Eliava appeared, they embraced like old friends.

At the end of his two-year term, Eliava was offered a permanent position as head of a microbiology institute in the south of France. "There are many microbiologists in France," he replied, "but I am the only one in Georgia. Georgia needs me." He began packing laboratory equipment that he planned to use to start a center of bacteriophage in Tbilisi—equipment given to him by his admiring colleagues at the institute. He could hardly have imagined that his idealism would doom him to a violent and premature death.

For the moment, d'Herelle resisted Eliava's pleas to come with him and help establish the center, though the notion of leaving Paris and starting over in a new place must have appealed to his keen wanderlust. In early 1921, he came back from an extended trip to Indochina, testing local phages against cholera, to discover that his laboratory at the Pasteur had been taken away from him. This was the doing of deputy director Albert Calmette, creator of the BCG (Bacillus Calmette-Guerin) vaccine for tuberculosis. Calmette had taken grave offense at d'Herelle's view that the vaccine would be useless in children. Jules Bordet, creator of the Bordetella vaccine, and Mihai Ciuca, a colleague of Bordet's in the Pasteur Institute's Brussels center, did some digging in the literature and found that in December 1915, a certain Frederick Twort, a British microbiologist, had published a paper on the same bacteria-eating microbes d'Herelle had noted two years later. Like d'Herelle, Twort had observed "glassy areas" on bacteria that indicated a zone of inhibition produced by something invisible. Like d'Herelle, he had theorized it might be a virus, but when he had failed to get funding to continue his research, he had let the matter drop. D'Herelle was the one who had coined the term *bacteriophage*. He had done more testing than Twort and put his findings in a broader context, much of which would prove amazingly prescient. Nevertheless, declared the Pastorians, Felix d'Herelle was a backdater or a plagiarist—or both.

D'Herelle was rightly outraged by the charge. Unfortunately, he chose to respond in the worst possible way. He could have acknowledged that another scientist had come to the same conclusion independently and managed to publish it first—a gesture that would have detracted minimally from d'Herelle's ongoing phage work and neatly disarmed his critics. Instead, he declared that Twort's finding was not related to bacteriophage and that perhaps an even earlier observation should be deemed the first. In 1896, Ernest Hanbury Hankin, a bacteriologist with the British colonial government in India, had noted that some waters from both the Jumna and the Ganges Rivers killed bacteria. Like d'Herelle, Hankin had passed this water through a type of filter that removed bacteria and noted that the water still demonstrated antibacterial action, so that something smaller than bacteria must be responsible. Hankin went so far as to hypothesize that this antibacterial substance was responsible for limiting the spread of cholera in India. But he had failed to follow up on his observation or try to categorize it in any way. Two years later, Russian microbiologist Nikolai Fedorovich Gamalea had made similar observations. D'Herelle's mention of Hankin and Gamalea seemed meant mostly to undercut Twort, but it was a strange defense. The stubborn Canadian added that his own early work with locusts—work that predated Twort's—had put him on the track to phages. But true as this may have been, d'Herelle had no published medical papers to prove it.

D'Herelle's claims strained the credulity of nearly all his colleagues, so in the fall of 1921, his relations with the Pasteur Institute poisoned, the embattled Canadian accepted an offer to do phage work at the University of Leiden in the Netherlands. He stayed for two years, during which he received an honorary medical degree, which he soon put to use. Hearing that a post in Egypt as head of a bacteriology laboratory for the League of Nations would soon need filling, d'Herelle sent his degree from Leiden to fulfill the requirement that he have a medical degree. He got the job, though not without tipping his hand to his critics back in Paris that he had no other degree to proffer. They noted, in particular, the time that d'Herelle had just spent in Indochina treating patients with phages. "If d'Herelle is not a physician," wrote his successor in that job to his chief critic, Bordet, "he has committed fraud."

In fact, d'Herelle had used phage therapeutically for years. In 1919, he and colleagues had used phages, almost certainly culled from human sewage, to treat a twelve-year-old boy stricken with severe dysentery. To assure that his concoction was safe, d'Herelle and colleagues first drank a goodly portion themselves—a noble choice quite at odds with his critics' characterization of him as irresponsible. Fortunately, nothing happened. They gave it to the boy, and after a single dose his symptoms disappeared. D'Herelle chose not to publish the results for several years, possibly for fear of being charged with medical fraud. In Indochina, he had used phage to treat the victims of an outbreak of bubonic plague—successfully, he reported. D'Herelle's biographer, William Summers, suggests that d'Herelle may have enlisted medical doctors to help him administer his phage—that like his hero, Pasteur, he was always able to coopt a willing young clinician to help him with his research on patients—but to his critics, d'Herelle's history of human experimentation was, at the least, ethically questionable.

One of the few who stood by d'Herelle through this period was Eliava. In addition to bacteriophages, the two now shared another passion: the ideals of socialism. Frequently before his return to Georgia in the fall of 1921, Eliava had dined with d'Herelle and raised toasts to Marx and Lenin. Later, Eliava's stepdaughter Hannah would recall that d'Herelle's bolshevism was confined to addressing men as "comrade" and refusing to kiss women's hands. A man wealthy enough to buy a car for his grandson's birthday, she would say, was not the sort of Communist she knew back home in Georgia. But nor, perhaps, was Eliava. Before leaving Paris, he reportedly bought several bottles of perfume for his great love, ballerina Amelia Vol-Levitskaya, and poured them into bacteriophage vials to smuggle the "bourgeois" goods past Communist customs officials.

How much of a role Georgia's native son Joseph Stalin played in establishing the institute's first home is unclear. But as the most powerful Georgian in the central government, Stalin must have known about Eliava's dream and at the least made it possible by choosing not to squelch it. To secure Stalin's blessing for what was to be called the Tiflis Bacteriological Institute, Eliava went through a local Communist leader he knew: Sergo Ordzhonikidze, the minister of light-heavy industry. In doing so, he slighted an old enemy: Lavrenti Beria, a rising star of Georgia's secret police, felt *he* should have been the conduit between Eliava and Stalin. For the moment, however, he could do nothing, since Eliava was clearly in Stalin's good graces. By 1925, Eliava was back at the Pasteur Institute, learning more and gathering additional equipment for Tbilisi.[2]

Having used phage with success in Egypt to treat Mecca-bound pilgrims stricken with bacterial infections, and then on victims of bubonic plague and cholera in India, d'Herelle in 1928 accepted a professorship in "protobiology" at Yale and started a phage center there. But his frequent departures began to offend his new colleagues, and squabbles over salary for the reduced time he was working led him, by the spring of 1933, to tender a letter of resignation. Another invitation came from George Eliava to join him in Tbilisi, and this time d'Herelle was ready to accept. A formal invitation from the Soviet government—one that apparently included assurances of a private house for d'Herelle and his wife on the grounds of the new All-Union Bacteriophage Institute, as it was to be called, and a car and driver—sealed the deal.

In his letters to d'Herelle, Eliava had made light of the growing political danger around him, but perhaps he was oblivious to it himself. His old nemesis, Beria, was now first secretary of the Georgian Communist Party and head of the Georgian secret police. Already Beria was showing signs of the extreme cruelty that would be his legacy, personally torturing prisoners and ordering his police to arrest attractive teenage girls so that he could rape them in his office. In 1930, he had ordered Eliava's arrest as a "wrecker," the term for those displaying antiparty activity. Only pressure from the central government, possibly from Stalin himself, had regained the microbiologist

---

[2]D'Herelle, despite his troubles, had just become much more famous, especially in the United States, with the publication of Sinclair Lewis's Pulitzer Prize–winning novel, *Arrowsmith*. Like d'Herelle, Dr. Martin Arrowsmith discovers that liquid in his laboratory flask formerly cloudy with bacteria is now mysteriously clear. "Something funny here," he says. "This culture was growing all right, and now it's committed suicide. Never heard of bugs doing that before. I've hit something! What caused it? Some chemical change? Something organic?" A colleague tells Martin he's not the discoverer after all. "D'Herelle of the Pasteur Institute has just now published in the Comptes Rendus, Academie des Sciences, a report—it is your X Principle, absolutely. Only he calls it 'bacteriophage.'"

his freedom. Even still, Eliava openly ridiculed Beria for his oafish ig-
norance of music, culture, history, and science. On at least one occa-
sion, he even mocked the party leader to his face. Summoned as a
leading medical expert in Georgia to draw blood from the arm of a
prominent local Communist Party official, Eliava learned only when
he arrived at party headquarters that the official was Beria. As Eliava
inserted the needle, Beria said jokingly, "Please don't take too much."
"Why not?" Eliava snapped. "You do from everyone else!" Almost
anyone could have guessed how all this would end.

D'Herelle and his wife arrived by ship in October 1933 from
Marseilles, to be met by Eliava at the port of Batumi. During this first
visit, d'Herelle did more than streak agar plates. Together, he and
Eliava used phage on a patient dying of a bloodstream infection
brought on by *S. aureus*. The patient's doctor made the request—
none of the feeble drugs he had at hand had worked—and the microbe hunters complied, albeit hesitantly, for phage remained an experimental therapy as far as the Communist government was
concerned. The patient had recovered by the time d'Herelle and his
wife headed back to the west in April 1934.

During d'Herelle's second Georgian stay, from November 1934 to
May 1935, he became so excited by the progress that he and Eliava
were making that he resolved to move to Tbilisi full-time as soon as
his personal compound was finished. During this second stay,
d'Herelle finished a monograph that Eliava promptly translated into
Russian. With a flourish, d'Herelle prefaced his work by writing,
"This book, which summarizes twenty years of seeking new pathways
in medicine, is dedicated to him who possesses ruthless and relentless
logic of history and who stands at the threshold of society. I dedicate
this book to him who has reached the top. I dedicate this book to
Comrade Stalin." Through his own connections, Eliava managed to
get a copy of the book to its dedicatee, and in so doing further in-
censed Beria.

By autumn, divisions in the Communist Party had become dangerous and visible. Stalin's Great Purge had not quite begun, but even an

ardent loyalist like d'Herelle now knew better than to return in the midst of such turmoil. For Eliava, the beginning of the end probably came in the summer of 1936, when an outbreak of dysentery occurred in eastern Georgia. Eliava oversaw a vaccination campaign in which two children who were treated died of anaphylactic shock. Beria's spies reported that Eliava had poisoned them. Eliava's greater mistake may have been to attract the attention of a woman sought after by Beria. Though married to Amelia Vol-Levitskaya, the prima ballerina to whom he had brought perfumes from France, he was an attractive man who liked to flirt. Whether this flirtation or a history of slights was the motive, Beria had Eliava arrested while he was sitting at dinner at home with his wife in January 1937. Several hours later, Amelia was arrested, too. Left alone in the house was Amelia's daughter by her first husband, Hannah, who was then about twenty years old.

Official documents list Eliava's date of execution as July 9, 1937, raising the possibility that he was tortured in the interim months, though other reports say he was shot the day of his arrest. Certainly, he had no trial. Hannah, unable to learn any information about him, stayed in her parents' apartment for three months, during which she married her boyfriend. In April, she too was arrested. She was thrown into a jail cell with a much older woman, very ill with swollen eyes, whom she realized with a shock was her mother. The two were separated as soon as guards realized they were related. Amelia died shortly afterward; Hannah remained in prison camps until Stalin's death in 1953, then returned to Tbilisi, her body battered but her spirit unbowed.

D'Herelle, focused so intently on his phages back in Paris, had no inkling his friend had been killed. For months afterward, he continued to send scientific equipment and letters to him. When he learned the tragic news at last, he was stunned. How could Stalin, his hero, have allowed this to happen? If d'Herelle rationalized the atrocity by blaming Beria and assuming the sadistic police chief had had Eliava shot without Stalin's knowledge, this conceit was soon dashed. Stalin summoned Beria to Moscow a year later to work as deputy to the chief of the Soviet secret police. Soon the chief had vanished, and Beria had taken his place.[3]

In the years since d'Herelle's first experiments with dysentery patients, a new industry of phage medicines had sprung up. An article in the *Journal of the American Medical Association* in 1933 reported that bacteriophages were "widely used for many types of bacterial infections" and noted that three well-known pharmaceutical companies were manufacturing them. But so little was actually known about phages that wildly exaggerated claims were made. One product called Enterofagos boasted miraculous powers against herpes infections and eczema—an unlikely combination, given that the two had nothing in common and neither was a bacterial infection. Others claimed to cure urticaria (an allergic reaction) and gallstones. As the inevitable failures mounted, phages began to look like snake oil.

Sadly, d'Herelle did nothing to clarify the debate. Stubborn as always, he insisted there was only one kind of phage—one kind that could adapt to kill all kinds of pathogenic bacteria. Worse, he declared in his 1922 book *The Bacteriophage, Its Role in Immunity* that "a single strain of the bacteriophage is rarely active for but a single species. Usually it attacks several species at one and the same time, and for each it possesses a separate and variable virulence." Initially, this was an easy mistake to make, since the average bucket of human sewage yielded so many different kinds of phages that one or another was likely to lyse whatever bacteria the distilled brew was used against. By the mid-1930s, however, any reasonable scientist researching phages as intensely as d'Herelle would have realized there were as many different kinds of phages as there were bacteria and that the whole trick was to match the phage to the pathogen it was designed by nature to attack.

---

[3]After Stalin's death in 1953, several Soviet leaders feared Beria might claw his way to the top, so Nikita Khrushchev had him arrested and tried as a spy in a secret trial. He was found guilty and was reportedly executed by a Soviet general after crawling on his knees to beg, weeping, for mercy.

In fairness to the beleaguered phage finder, there was a lot about phages that no one could know at that time—knowledge that directly affected how well they worked. Early preparations were often impure: they contained bacterial debris from cells already lysed by the phages, even tiny amounts of which could cause endotoxic shock. Today such debris is assiduously removed by pharmaceutical companies from any prospective drug; in d'Herelle's time, no one appreciated the problem. Adverse effects did lead manufacturers to try to sterilize their preparations, but they did so by heating them, which cooked the phages like hard-boiled eggs and left them useless, or by adding mercurials and oxidizing agents, which also inactivated them. Along with not identifying which phages worked against which bacteria, neither manufacturers nor scientists yet knew that some phages were lytic and others were temperate.

In the West, the new phage industry was all but obliterated by World War II. With the fall of France, the Pasteur Institute was occupied by the Nazis. Apparently the Nazis tried, without success, to gear up large-scale production of phages for treating battlefield infections. Hitler reportedly had a plan code-named "Edelweiss" to cross the Caucasian mountains and seize the Georgian oil fields, as well as the Eliava Institute's phages for treating battlefield infections, but the troops he sent were turned back by Georgian partisans.

The Russian army did carry phages into war, first against Finland in 1938–39 and then in World War II, using them both as therapy and as preventative medicine against dysentery and gangrene. And in Tbilisi, where the production of phages was coordinated with new centers in Kiev and Kharkov, various studies survive to attest to the contribution phages made on the battlefield. But the Soviet use of phages remained all but unknown, and unevaluated, in western Europe and the United States, and had this wartime experience with phages reached the West, few if any doctors would have taken it seriously. No Soviet battlefield doctors, after all, were engaging in proper control studies, withholding phages from some soldiers while giving it to others. And the products of a small, quirky institute in Soviet Georgia would hardly have been tested, much less used, by doctors in England and the United States. Who in his right mind would muck about with human sewage as therapy after the arrival of penicillin in 1943?

Despite d'Herelle's intransigence (which ended only with his death in 1949), the grievous mishandling of phages, and the coming of penicillin and the antibiotic age, a fascination with phages persisted behind the Iron Curtain, unrecognized by the West. In Russia, as in the United States, antibiotics were miracle drugs. But phages had helped save soldiers' lives. And World War II had shown the Soviet government that phages caused no side effects. (The only problem had been occasional liver pain attributed to the release of endotoxins when bacteria were destroyed.) To the researchers keeping alive Eliava's flame at the Tbilisi Institute, there was even a case to be made that phages were *better* than antibiotics. No drug could multiply as phages did, for example. All a patient needed was a dose of the right phage—by now, the researchers at Tbilisi were well aware of the need to match specific phages and bacteria—and the little viruses would replicate to become, in a matter of hours, a population of millions. The introduction of a mere 100 phages into an infected wound was enough to destroy, in short order, 100 million bacteria. When the infection was essentially destroyed, the phages just washed harmlessly out of the patient's body. And no drug could concentrate itself at the site of an infection as phages did.

As resistance became an issue with antibiotics, the Soviets also noticed that phages, while incurring some resistance, provoked a whole lot less of it. In the rare cases in which it did occur, there was no need to spend millions of dollars devising a new antibiotic. New strains of phage that lysed the bacteria could be found in days or weeks: all one had to do was resort to the nearest repository of human sewage.

Throughout the next three decades, scores of medical papers were generated about phages by the Tbilisi researchers. Crucial work was also done and published, particularly in using phages to treat drug-resistant bacteria, by a team led by Dr. Stefan Slopek at the Polish Academy of Sciences in Wroclaw: an overall 92 percent rate of success was achieved with 550 patients, aged one week to eighty-six years, suffering from infections resistant to all available antibiotics. But these papers, and phages in general, were ignored.

With one exception. Beginning in 1982, British researchers Williams Smith and Michael Huggins studied the effects of phages on animals. They collected phages from human sewage stations, animal markets, and farms—basically anywhere, as Huggins put it, "where you got a collection of shit." They cured mice infected with deadly strains of *E. coli,* tested phage successfully on calves, lambs, and pigs, and showed that with farm animals, phages could be transferred from animal to animal without human intervention—a contagious cure. Huggins reported later that when sick cows under study were treated with phages, their diarrhea cleared up immediately. The efficacy rate, he said, was 100 percent. The work of Smith and Huggins, and then of British colleague James Soothill, who used phages to cure guinea pigs and mice infected with *Pseudomonas,* alerted a few microbiologists, Glenn Morris among them, to the idea that phages deserved another look. But these researchers remained few, and they were usually dismissed.

At the Eliava Institute, the West's disinterest in phages mattered not at all. To researchers who had worked with George Eliava and never left, the 1980s were a golden period. Phages were manufactured in tablet and liquid form and sold all over the Soviet Union. Soviet troops used phages prophylactically, to prevent gangrene. (Most recently they carried them into war with the breakaway republic of Chechnya.) Children in pediatric hospitals were given phages to prevent the spread of *Salmonella* and other gastrointestinal problems, including *Shigella* dysentery. Phages were given for *Pseudomonas* infections in cystic fibrosis patients, urinary tract infections, and staph infections. For staph infections of the skin, powdered phages were applied. For systemic staph infections, a special phage remedy was administered intravenously (by infusion, transfusion, and injection).[4] Just from the institute's own plant, two tons' worth of phages were produced daily, much of it sent beyond Georgia's borders. Other plants, scattered around the Soviet Union but coordinated with

[4]*Staphylococcus* phage was mostly used for treatment of chronic septicemia, for treatment and prevention of eye, ear, throat, and lung diseases, for healing burn wounds, and for infections arising from surgery.

Tbilisi, made more. All these products generated healthy profits and supported a staff of as many as 700 researchers and technicians, who worked at the institute as sheltered from the deprivations of a collapsing political system as any Georgians could possibly be.

Then, in 1989, Communism began to crumble. In Georgia, long-suppressed cries for independence brought Soviet troops, then anarchy. Within months, the 700 researchers had been reduced to 200. To those left holed up in the institute as bullets flew downtown, the legacy of George Eliava—and a unique library of thousands of phages kept active in refrigerated storage for as long as seventy years; a vast repository of biological knowledge to help fight new bacterial strains—seemed on the verge of destruction. Not one of those 200 beleaguered researchers, in their wildest imaginings, would have predicted that a new, even more promising age for the institute and bacteriophage was about to begin.

Each morning, Dr. Nina Chanishvili and her colleagues would hurry from their homes to the institute in groups of three or more, walking because public transportation had ceased. In the distance, they could always hear gunfire, and everyone knew someone who had been killed by a random bullet. There was no guard at the institute's doors; the researchers had to hope that Georgia's clashing private armies would see no point in appropriating vials of phages and bacteria. Three evenings a week between 6:00 and 10:00 P.M., Chanishvili taught a biology course at a private college. She had to walk alone to the college, and then home, as gunfire pierced the night. That scared her the most, but she had no choice: with the institute's budget cut back to almost nothing, the teaching job was her main source of income.

One day at the Eliava, the electricity failed. Over the next months, it went off more and more often, until in 1993 it stopped coming on at all, a function of government indifference toward the institute. The researchers packed their home refrigerators with phages; those had power, at least, a few hours a day. One night that year, the heat went off, too, and simply stayed off—for years.

Russia had cut off more than heat and electricity. It had cut off all business ties. There were no more orders for tons of phages, nor was vital laboratory equipment sent, even at a price, from Moscow. Devastated, the institute's directors—Nina's uncle, Teimuraz Chanishvili, and another devotee of decades, Amiran Meipariani— cut staff to a bare few researchers to try to keep the phage library alive. Salaries were slashed to the absurd. In the worst days of 1990, Nina earned the equivalent, in rubles, of fifty cents U.S.—a month.

Without light and without heat, Chanishvili and the other researchers worked by natural light, bundled in their winter coats much of the year, typing with their gloves on. To warm themselves, they used portable kerosene heaters and burned tablets of dry alcohol to heat coffee or tea: two tablets per cup. In winter, at least, the phages could be kept alive without refrigeration because the ambient temperature was so cold. In the summer months of the early 1990s, with electricity intermittent or completely off, nearly half of the institute's phages died—some 500 clones that worked against one or another strain of about a dozen bacterial species.

A first ray of hope came indirectly from George Soros, the currency-trading billionaire who had seeded various foundations in eastern Europe to promote democracy with the end of the Cold War. In 1992, Soros's Tbilisi foundation gave the institute a one-time $30,000 grant to stave off complete collapse. The grant bought labware and computer parts and paid for shoestring salaries and overhead, though it raised expectations the foundation chose not to meet. The researchers were disappointed that Soros himself never visited the institute and that no follow-up grants materialized.

For Chanishvili and her colleagues, the period of 1993–94 was the worst time. The civil strife had ebbed, but supplies were dangerously scarce, with hardly any electricity and no heat at all. At the institute, more phages died, while a staff of twenty struggled to keep the place from closing altogether. Only through friends in Switzerland was Chanishvili able to buy a sack of flour, a sack of sugar, and a sack of potatoes, which kept her family alive.

The institute's best bet was to try to interest some European or U.S.

medical entrepreneurs in investing in phages for profit. But the directors were clueless about how to do this. They were wary, too: decades of Communist-bred paranoia could hardly be dissipated overnight. Unbeknownst to them, however, a senior scientist at the National Institutes of Health in the United States had been fascinated by phages for years and had just conducted a remarkable experiment that seemed to solve one of the biggest drawbacks to their therapeutic use. So exciting were the results that they were leading to a first phage start-up in the United States—and the start of a new era, in the most scientifically advanced country in the world, for a bacteria fighter not taken seriously for more than fifty years.

Carl Merril had learned of bacteriophages as most molecular biologists of the mid- to late twentieth century had. They were viruses that could be manipulated to move genes into bacteria: a critical tool of molecular genetics. In 1945, Salvador Luria and Max Delbruck had studied how a phage takes over a bacterium's replicating machinery and gets it to make phage instead. Why did some bacteria resist the takeover? Luria and Delbruck proved that spontaneous genetic mutations were responsible. Then Luria and Alfred Day Hershey had shown that phages mutate, too, often in response to bacterial mutations: a kind of constant one-upmanship. Separately, Hershey and Martha Chase had shown that DNA was the substance that phage insert into bacteria to get bacteria to make more phages. Together, these findings had shown the utility of phages in moving genes into bacteria and also led directly to James Watson and Francis Crick's 1953 elucidation of the structure of DNA as the "double helix," laden with genes, that was the essence of biological life.

By the time Carl Merril arrived at the National Institutes of Health in the late 1960s as a research physician, early efforts were under way to use phages to move all kinds of genes. The field was so new that no one knew if these viruses were safe to work around. Merril made it one of his missions at the NIH to determine how safe phages really were.

The somewhat anticlimactic answer was that most phages were very safe indeed and the few bad apples could easily be avoided. But as part of his due diligence on the subject, Merril determined that phages *could* sometimes carry genes into mammalian cells, including human ones, not just bacterial cells, though the phages could not multiply there. This was a shock to his colleagues and a discovery that helped make possible the future miracle of cloning. These gene carriers were the phages called temperate—the ones that colonized their host cells rather than destroying them. But Merril's work also exposed him to the whole colorful history of lytic, or cell-bursting, phages as human therapy. He became fascinated by the stories of d'Herelle and Eliava, of the rise and early fall of "bacteriophages." He came to see the discrediting of phages for human therapy as a great and unnecessary tragedy—perhaps, he thought, one which he might help undo.

The Georgians, from what Merril could see, administered phage topically or orally. Topical phages might help skin infections, and oral phages might work against gastrointestinal pathogens. But unless you injected them, phages wouldn't work against serious bloodstream infections. The Georgians *said* they injected phages and got good results, but Merril doubted that. What he found was that when he injected phages, they were quickly taken out of circulation by the body before they could reach the site of bacterial infection. Still, this purging was curious. If the mice he was working with had been injected with phages before, the reason would be obvious: their immune systems had formed antibodies. But the phages seemed to be new foreign agents. The formation of antibodies to fight them would take days or a week. What, then, was sweeping them out so quickly? And was there any way to circumvent that force?

The NIH had germ-free mice, birthed by cesarean section and grown in sterile chambers. No germs meant they'd developed no antibodies. Merril injected them with *E. coli*, then with *E. coli* phages to see what would happen. To his surprise, the phages were still prevented from reaching the *E. coli* bacteria. They were entrapped, Merril saw, by the animals' reticuloendothelial system: the liver and spleen.

Other kinds of viruses, of course, had learned how to evade the liver and spleen: hence the spread of all manner of diseases, from AIDS to the common cold. Perhaps, Merril theorized, phages had remained vulnerable to the liver and spleen because the bloodstream of mammals was new terrain for them. They had existed in soil and in water; in mammals, they had been parasites to skin bacteria, and to bacteria of the gastrointestinal tract. But organisms in the GI tract were not, strictly speaking, *inside* the body. They had no access to the bloodstream, unless the tract was leaky. Aside from their short stint as human-bacteria eaters in d'Herelle's time, and locally since then in eastern Europe, phages hadn't ever been called on to infiltrate the mammalian circulatory system. So, in Merril's view, they hadn't ever figured out how to resist the sequestering pull of the liver and spleen, as other viruses had done.

Merril published his findings in 1973 and more or less forgot about them for twenty years. Then, as he read more and more articles about antibiotic resistance, he thought again of his early experiment. If phages could be kept from being expelled by the liver and spleen so that they could reach the site of infection and start multiplying exponentially, might they not be a brilliant alternative for doctors running out of antibiotic choices? Now, as chief of the Laboratory of Biochemical Genetics at NIH, Merril met Richard Carlton, a New York–based psychiatrist involved with technology transfer. Carlton had been working with the researchers of a potential new drug to combat septic shock. He became excited by Merril's talk of his early phage experiments and urged the NIH man to do more. This time, Merril and an NIH colleague, Sankar Adhya, injected a large number of phages into an animal and noted that within an hour most of the phages were washed out. When they drew a syringe of blood from the animal and put it in a petri dish, however, they could see that some phages were still swimming around in it. They knew this because when bacteria were added to the blood, small circles of clearness showed still-active phages inhibiting the bacterial cells. Each of these "plaques," as the clear circles were called, now consisted of many more daughter phages. Merril thought these might be mutants, phages hardier than the rest. He injected the mutants into a second animal and culled the survivors of that cycle. He did this about ten times. At the end, he had what appeared to be a nearly pure culture of tough, scrappy mutants, all capable of withstanding the excretory force of the liver and spleen. Merril, Adhya, and Carlton called the process "serial passage" and together won a patent for it on behalf of the government.

With that, Carlton staked $50,000 of his own money to start a company he called Exponential Biotherapies. Merril and Adhya began working with him through a Cooperative Research and Development Agreement by which government scientists and private companies were able to pool their efforts. The trio ran more experiments and watched serial passage occur in an entirely consistent manner, just as they had predicted. After the first generation, only one phage of every 100,000 injected survived. After the eighth passage, the number of survivors had increased 63,000-fold. These superphages would still be eliminated after forty-eight hours, when the body finally developed antibodies. And they might never work again in that individual patient, presuming those antibodies remained part of the immune system, capable of mounting a much faster attack the next time. While they *were* in the body, however, phages could be more effective than antibiotics, or so Carlton and Merril believed. Ten to a hundred molecules of an antibiotic were generally needed to kill one bacterium: the molecules had to gang up on the cell. A single phage killed a bacterial cell and created 200 daughters in the process. Less than an hour later, when those 200 daughters each attacked 200 more bacterial cells, they propagated 40,000 daughters. "You've been spoiled, Richard," Merril said when he called Carlton with the results. "Most science doesn't work this well." Merril and Adhya, urged on by Carlton, began nurturing serial-passaged phages to hit one of the most important drug-resistant targets in the United States: vancomycin-resistant *E. faecium*.

Merril, Adhya, and Carlton published their findings on serial passage, after the inevitable academic lag time, in the April 1996 issue of the august *Proceedings of the National Academy of Sciences*. This in itself was a triumph. In the same issue, the authors got a tremendous boost from one of the most distinguished scientists in the world. Joshua Lederberg, former president of Rockefeller University and a Nobel laureate, had shown, with Edward L. Tatum in 1946, that bacteria conjugate, and later, in 1952 with Norton Zinder, that phages can transfer DNA from one bacterium to another. In a signed commentary, Lederberg praised Merril's "elegantly simple" experiment and wrote that it was "an ingenious surmounting of one of the hurdles to the use of phage in therapy." Lederberg worried that bacteria would likely mutate to resist phages, but these mutations, he said, "might be delayed empirically by the use of a cocktail of phages with different bacterial receptors. In these days of genetic engineering, other tricks come to mind." At any rate, Lederberg wrote, "Along with the excitement of a new approach to antibacterial therapy, at a time when many antibiotics have run out, should be a renaissance of study of bacteriophages ... a subject scarcely mentioned since d'Herelle's time." Having Lederberg take phage seriously was like having Saul Bellow write a rave review of an unknown writer's new novel, or Robert Redford champion a newcomer's film at the Sundance Film Festival. Overnight, it put phage on the map.

The *Proceedings of the National Academy of Science* was hardly a staple of venture capitalist Caisey Harlingten's airplane reading. But as he sat beside his girlfriend on a business flight, he started reading an article on phage in the November 1996 issue of *Discover* magazine that excited him far more than Lederberg's commentary would have done a few months before. "The Good Virus," by medical writer Peter Radetsky, suggested that d'Herelle's discredited discovery might be due for a dusting off. Two respected scientists, geneticist Jim Bull of the University of Texas at Austin and Bruce Levin, a population biologist at Emory University in Atlanta, had decided to rerun the 1980s phage experiments of Englishmen Smith and Huggins on animals, using strict controls, and see what they found. To their surprise, Radetsky reported, phage had saved far more mice from lethal

E. coli than streptomycin had. Radetsky went on to explain the work being done at the Eliava Institute and to quote a doctor from the Georgian State Pediatric Hospital as saying that his hospital had had such success in using phage to treat drug-resistant bacterial infections that the doctors now gave phage as a preventative therapy to every incoming child. Unfortunately, Radetsky reported, while the Eliava Institute was still producing phage, and interest in phage was growing, the institute's very walls were crumbling. "We are in a miserable position," Nina Chanishvili was quoted as saying. "Today the institute has but half a life—but still it is existing."

Harlingten could barely suppress his excitement when he finished the article. He knew little about science, but he knew an opportunity when he saw one. He had made millions by licensing cutting-edge technology, including a "blue laser" smaller and more efficient than earlier lasers because it operated without heat, and a "virtual vision" display system in which computer- or camera-generated imagery could be encoded on a very low wattage laser and scanned directly onto the retina of the human eye, without the need for a screen. Incredibly, Harlingten had learned about both technologies from articles in *Discover* magazine. Here, clearly, was lightning striking a third time. And here, as Harlingten told his girlfriend, was one of those rare, golden chances to do well by doing good.

Within a month, Harlingten flew over to Tbilisi to introduce himself and get a tour. Dilapidated though the institute was, the Eliava's directors assured him only $2 million would be needed to spruce it up and equip it with vital new laboratory equipment. Harlingten soon raised $3 million, about 25 percent of which represented his own investment, to create a business registered in the United States but based on work to be done at a prospective new lab on the grounds of the Eliava Institute. In honor of his new partners, Harlingten called the venture Georgia Research Institute and doled out a small portion of the money—about $28,000, according to Nina Chanishvili—to seed the new lab with new equipment. What he needed next, he felt, was a top U.S. lab to work with the Tbilisi team and start the tricky business of turning local Georgian phages, presumably from human sewage, into a product that the FDA could approve for sale in the United States. After asking around, he called Glenn Morris in Baltimore.

Ever since VRE had become a common visitor to his VA medical center in about 1990, Morris had cast about for help. In his university lab, he and his team had tested all possible combinations of other antibiotics against their hospital VRE strains. None worked, either on its own or in combination with others. In desperation, Morris began reading the scant literature he could find on phages—an alternative suggested by a research fellow in his lab named Sandro Sulakvelidze, who just happened to have come from the Republic of Georgia. The literature was, as he put it, lousy. The only phage studies on humans had been done east of the Iron Curtain, mostly in Poland and Georgia, with poor controls and no molecular biology at all. Basically, the eastern Europeans had just given phages—of what chemical composition they never said—to a lot of people with infections of one sort or another and recorded how many were "cured." The percentages were intriguingly high, but there were no control groups for comparison, and no follow-up studies.

Still, Morris felt in no position to scoff at phage as so many of his academic colleagues did. As a medical doctor, he now saw firsthand the debilitating effects of VRE. If phages worked in Tbilisi, perhaps they would work in Baltimore. Together with Sulakvelidze, Morris helped the Eliava Institute apply for U.S. government funds to refine a phage for VRE. To their chagrin, the proposal was denied. Right after that came Harlingten's call.

It seemed nothing less than providential. Gushing with enthusiasm, the West Coast venture capitalist offered funding for joint research between Morris's lab and his own new lab at the Eliava Institute—essentially to fund what the scientists had hoped the government would sponsor. Eagerly, the two sides got to work. From Baltimore to Tbilisi went hospital strains of various infectious bacteria, including many MRSA and VRE strains. Eventually the

Georgians would use an intravenous phage to kill 84 percent of the 93 MRSA strains they tested from Morris and 97.8 percent of their 186 enterococcus strains.

Meanwhile, Morris and Sulakvelidze began gathering phages in the Baltimore area, which was to say they went mucking about in the university sewage system, Baltimore's harbor, and the Chesapeake River, taking samples of water swimming with fecal bacteria—and phage. Every time they found a phage that knocked out a strain of VRE, they added it to their growing phage library. Morris also sifted his water samples for phage to combat some of the more than 500 strains of *Salmonella* he had in his lab, most of them from FOODNET, the national infrastructure of testing sites for food-borne bacterial disease that he had helped establish as an outside consultant to the Clinton administration—the network that Fred Angulo now ran. Morris found a lot of phages that knocked out one strain or another. Even more heartening, he found that phage "cocktails" of seven or eight phages could kill 95 percent of his *Salmonella* strains.

As phages and bacteria flew from Baltimore to Georgia and back, Caisey Harlingten decided he needed a CEO to take daily charge of his phage venture. He chose Richard Honour, a wry, white-haired West Coast veteran of technological start-ups in fields ranging from biotech to biological warfare. Honour reviewed all of Harlingten's business documents and articles about phage, then flew up to meet the founder and a few of his investors for a bonding weekend of talk and fun at the Whistler Ski Resort in British Columbia.

"I've made two conclusions," Honour cheerfully told the group in one of the resort's conference rooms. "One is that phages are excellent candidates to address the issues of antibiotic resistance." The group exchanged self-satisfied grins. "The other," said Honour, "is that you need to reconceive the company completely."

Honour explained that he'd just spent four years helping a company try to win FDA approval for Russian-grown peptides. The founders had raised $28 million and had had a far easier product to pitch than phage, "and they still didn't get the license." There was no way that the FDA was ever going to grant U.S. approval to natural products gleaned from Russian sewage. Products, moreover, that were actual living organisms, *viruses* that proliferated exponentially in the human body and mutated into God knew what. Could Harlingten even imagine the challenge of persuading a pharmaceutical plant to run mutating viruses through its machinery? Viruses that might somehow contaminate all of its other products?

There was a stunned silence in the room.

"Okay," Harlingten said at last. "Let's go to Georgia." Honour would see the institute for himself and ask all the tough questions he wished of the scientists involved.

In Tbilisi, Honour had to jump over bomb craters to cross the streets. The institute's labs still had no electricity or heat. And the phages! God-awful mixtures, Honour muttered. Pus, urine, and excrement from sewage—the very antithesis of what physicians had been preaching since Pasteur proved his germ theory. Honour was given a tour of a few almost presentable labs in the building. As he left, he was sure he saw researchers disassembling them—a Potemkin project laboratory, in effect. As for the manufacturing plant, it was like a long-abandoned auto plant from before World War II, with rusted, inoperable, hopelessly outdated machinery. What the Georgians really needed, he realized, was at least $80 million from their new angel, Caisey Harlingten, to reestablish their infrastructure. And even then they'd have a product, he felt, that the FDA would never approve.

Perhaps, thought Honour later, he should have just walked away. But there *was* something awfully compelling about phages. They did work, at least some of the time. In the Soviet Union over the last seventy-five years, Honour reckoned, they had probably saved millions of lives. And now the creeping crisis of bacterial resistance, both in the West and the East, offered what any business person could see was a wide-open market need, a great global opportunity. But if phages were to stand any chance of winning FDA approval, Honour told Harlingten, there was an obvious, unavoidable step the start-up had to take.

"What's that?" Harlingten said.

"Cut all ties to Georgia," Honour declared flatly.

Harlingten blanched. In his first visit to the institute, he had forged strong personal bonds with the Chanishvilis, Teimuraz and Nina, and with co-director Amiran Meipariani. But Honour was right. There was no need to further the science in Georgia, no point at all. The same business could be started in the West at any educational institution, or under private auspices.

"I'll be the bad guy," Honour said. "I'll break the news to the Georgians. And then we'll go back to the U.S., and we'll find phages there." Under a new name, he explained, the reconfigured start-up would then genetically modify the best phage it found.

There was a pause.

"How long will that take?" Harlingten asked.

"Maybe seven years."

That year, the Americans severed all business ties with the Eliava Institute. They consolidated their company in Seattle, Washington, and began testing U.S. phages against methicillin-resistant *S. aureus,* sensitive and resistant forms of tuberculosis, *Pseudomonas aeruginosa,* and many of the other usual suspects. They also changed the name of their company, from Georgian Research Incorporated to Phage Therapeutics. "We gave the Americans access to all this background research," a bitter Nina Chanishvili later told the *New York Times,* "and they simply walked away with it. They told us we were stupid at business. Well, that at least was true."

In Baltimore, Glenn Morris and Sandro Sulakvelidze watched with dismay as Caisey Harlingten cut loose from Tbilisi. They felt the Georgians had not been treated well. They were sure, too, that Richard Honour was wrong to pursue genetically engineered phage. Indeed, the very notion was oxymoronic. If phages were all made of nucleic acid, and nucleic acid naturally mutated over time, an "engineered" phage would do the same, and thus be, in a sense, a living organism. And in any event, why try to improve on what already worked so well in nature?

After bitter words, the Baltimore contingent washed its hands of

the rechristened Phage Therapeutics and started talks with Richard
Carlton of Exponential Biotherapies. Soon, however, this potential alliance fizzled, too, and the Baltimore group broke off talks. Instead,
they began talking about forming America's third phage start-up.

Now Morris and Sulakvelidze made their own pilgrimage to
Tbilisi, meeting Nina Chanishvili and the other researchers. Morris
told them they were interested in forming a U.S.-based phage company but one with strong links to Tbilisi. He saw a business opportunity, but also an opportunity to do an enormous amount of good. For
starters, Morris said, he wanted a phage to treat active VRE infections, but he also wanted to see if he and Sulakvelidze could develop
a prophylactic VRE phage: one to be given to the 30 to 40 percent of
his hospital patients who were merely *colonized* with VRE. If he could
keep the colonized patients from becoming infected, the battle would
be more than half won.

As an old pro, Morris knew all the medical establishment's objec-
tions to phage. True, phages were made of protein and nucleic acid
foreign to the human body and so would be targeted by the immune
system eventually. But if you chose the right phage, the right dosage,
and the right means of administering the virus, perhaps it could get to
the site of an infection before the immune system created antibodies
for it. By the time the antibodies arrived, the phage would have destroyed all the pathogenic bacteria. As for Carl Merril's warnings
about the liver and spleen removing phage too quickly without his
trick of "serial passage," the Baltimore contingent just didn't see the
problem. Why, after all, did phage work with such astounding success in all those eastern European studies?[5]

The more legitimate concern, Morris felt, was that bacteria might
grow resistant to phage. But the beauty of natural phage as opposed to
the genetically engineered stuff that Richard Honour was proposing to
make was that it could mutate right along with the bacteria. This was

---

[5]Merril and Carlton agreed that phage worked in such studies—but at high and inefficient
doses. Serial passage, they felt, allowed for lower doses of more concentrated phage, with a pro-
portionately lower antibody response, so the phage would be cleared from the body more slowly.

what phage had done with bacteria through the whole of geologic time. Then, too, as Joshua Lederberg had observed, "cocktails" of half a dozen or more phage could hit bacteria from all sides, making resistance—in the short run, at least—almost mathematically impossible. And when resistance did occur, the Eliava Institute could come up with a new phage that worked, at almost no cost, in a matter of weeks.

The greatest problem, as Morris saw it, was specificity. One did have to match the right phage to the right strain of bacteria. And so for patients who needed immediate, life-saving help, phage would be inappropriate. It was just too narrow a spectrum to guess about before you had a culture analysis. But patients who could wait a couple days for their blood results to come back from the lab might do far better with the right, matched phage than with a broad-range antibiotic. Most antibiotics, after all, killed a lot of "good" bacteria along with the pathogenic strains and often caused side effects, particularly *Clostridium difficile* diarrhea, the so-called antibiotic diarrhea, while phages caused almost none.

The fact was, Morris would observe to his skeptical colleagues, that medical science in the last years of the twentieth century relied 100 percent on one kind of bacteria fighter: antibiotics. And he hardly had to tell his colleagues what was happening to them. Phage might not be the new panacea, but it might help. Of course, so would new vaccines. And so might immunotherapy—injecting human antibodies as direct therapy against an infection. By cycling from one approach to the next, doctors could keep bacteria off balance and so stave off the time of inevitable, all-out resistance to each new drug— or, for that matter, each phage. *You have to think outside the box,* Morris felt. For her part, Chanishvili felt that Morris was a trustworthy sort, and she was reassured by the involvement of fellow Georgian Sulakvelidze. They were all willing to take a chance.

Out in Seattle, the self-appointed bad guy of Phage Therapeutics began working toward an all-powerful, genetically modified phage against drug-resistant *S. aureus*. Richard Honour's researchers collected phages from all over North America and much of Europe: the whole prospective market for their product. By mid-1999, he had twelve genetically derived phages that worked very well against staph, or so he said. Of the twelve, three were superb and one was the best. Honour called it Rambo because, as he said, "it takes no prisoners and it kills everybody." Honour told his investors that Rambo routinely killed 93 to 95 percent of all *S. aureus* isolates, no matter how resistant, for patients all over the world. "And the best thing is," he said, "we can manufacture it again and again."

In early September 1999, Honour's researchers were still injecting mice with Rambo—a year away, at least, from human clinical trials— when Honour got a desperate call from a man in Hamilton, Canada, who had just seen a Canadian broadcast of a BBC show done on Phage Therapeutics and the Georgia connection. The man explained that his mother had come in for surgery to correct a life-threatening heart condition known as Marfan syndrome, in which the walls of her aorta disintegrated. The woman had had her aorta replaced with a prosthetic device but then developed a systemic multidrug-resistant *S. aureus* infection, probably from bacteria on the device itself. Her doctors had tried every antibiotic they could think of, but her condition had rapidly declined. Now she was on life support. The doctors had one last strategy: to replace the infected prosthetic device with an aorta from a cadaver. But they saw no point in trying that unless they could rid her bloodstream of the *S. aureus* infection. The woman's son had heard Honour and the Eliava researchers on the television documentary describe how miraculous phages were against multidrug-resistant *S. aureus*. What could he do to get Honour to send him some right away?

"I don't have an approved product," Honour explained, gesturing as he talked to the executives gathered in his office for a strategy meeting. "But I don't want your mother to die. So here's what we have to do."

Honour told the son to seek permission, on a strictly confidential basis, from all the hospital's authorities: not just the woman's doctors, but the human ethics committee, the clinical microbiologist, the administration—every echelon of the institution. All had to authorize this unapproved therapy on a compassionate-use basis. If they all agreed, then the woman's doctor should FedEx to Honour a clinical isolate of her blood.

The man did as Honour advised, and the next day a vial of the woman's blood arrived at the offices of Phage Therapeutics. In the company's lab, researchers set the multidrug-resistant *S. aureus* bacteria against "Rambo."

The phage obliterated the bacteria.

By FedEx that day, Honour sent back two doses of Rambo to Hamilton. The vials were picked up by a hospital employee from the Toronto airport, rushed through customs, and driven immediately to the hospital. When the woman's chest was opened up that day, one of the surgeons kept Honour on an open phone line so the entrepreneur could offer his thoughts on how to spray the phage directly onto the woman's infected heart. For good measure, the doctors injected the second dose into her veins.

The phage acted even more impressively than Honour had hoped. First, it exploded the bacterial cells of the infection. Then it sent the bacterial debris of those cells through the woman's blood, provoking a strong response from her immune system. The immune response spiked a temporary high temperature. It also eradicated her concurrent urinary tract infection. In less than twenty-four hours, the woman was free of *all* infection.

Despite a pact of confidentiality between the hospital and Phage Therapeutics, word of the curious drama spread, and newspaper reporters began peppering the hospital for details. The hospital authorities grew exasperated and blamed Honour for publicizing the incident to promote his product. Canada's Health Protection Branch then threatened to arrest Honour for transmitting an unlicensed medicine across the Canadian border. Worst, the woman died after all. But not for many months, and not, according to Honour, from multidrug-resistant *S. aureus:* free of infection, he claimed, her debilitated heart simply gave out in the end.[6]

Unfortunate as its denouement was, this first case in the West

[6]The hospital would not comment on the case, but one unsubstantiated report, secondhand from a doctor attending the woman, suggested the infection *had* returned and that the woman had died from it after all.

stirred enormous interest in the field. Doubtless, it also helped Honour to raise $3 million in a second round of financing and to begin to define his targets for human clinical trials. First, he wanted to set Rambo against staph infections of the eye—a localized and so fairly containable condition. From there he would take on staph infections of the skin, then infections of the lungs and urinary tract, and finally the hardest challenge, bloodstream infections. This was a cautious approach, but it did, at least, focus on human infection. You could win approvals sooner and get out a profit-making product more quickly if you started by lowering your sights to plant and animal infections. But that, Honour implied, was farm-league stuff, literally and figuratively. If his competition wanted to mess with that, he was more than happy to let them.

That, indeed, was exactly where Glenn Morris's Intralytix was messing by the fall of 2001. After four years of trading isolates with Tbilisi, the Baltimore start-up was in negotiations with the NIH to sponsor phase one safety trials on a VRE phage. According to Morris, a first round of financing would soon yield $8 to 10 million in development funds. At the same time, with the help of another microbiologist over from Georgia, Zemphira Alavidze, Intralytix had identified phages that killed several strains of methicillin-resistant *S. aureus* isolated from hospital patients across the street from Morris's lab. But in his new role as entrepreneur, Morris now appreciated how long a gamut he faced with the FDA to get phages approved for multidrug-resistant infections. And while his *S. aureus* phage was killing infections in the lab, the patients who had those infections were dying across the street. Reluctantly, Morris had come to see that the fastest way to get a product to market was to try phage on farm animal infections.

The obvious target was *Salmonella*. Morris had helped put in place new federal food-safety regulations, just now swinging into effect, that used *Salmonella* as their benchmark pathogen. For meat producers struggling to keep their *Salmonella* levels down while their antibiotics grew less and less effective, phage might be the perfect solution. And as Morris had proved in the lab, he and Sulakvelidze had phage cocktails capable of killing most of the more than 500 strains of *Salmonella* they had on hand.

They began focusing on poultry, where *Salmonella* levels were shockingly high—a problem ameliorated only by consumers' knowing to cook chicken thoroughly, which usually eradicated any chance of direct infection by ingestion. With antibiotics, producers could contain the problem but never eliminate it, because eggs were hatched with *Salmonella* on them and antibiotics used at that point affected the hatching rate. But phages, it turned out, could be sprayed on eggs without harm. They could be used on freshly hatched chicks, too. Finally, they could be sprayed on a "post-chill" bird: a chicken killed, processed, and about to be packaged. Phage for fowl: it even had a catchy ring to it. But the achievement of even this modest goal seemed many months away.

Of the three start-ups, Exponential Biotherapies in early 2002 seemed in some ways a dark horse. Richard Carlton's company lacked a genetically engineered "Rambo," allegedly successful in one direly sick hospital patient, that Phage Therapeutics had. Nor did it have a pipeline to the Eliava Institute, still the world's largest brain trust on phages, that Intralytix did. But Carlton claimed to have raised $15 million since starting the company and he had completed phase one trials (on healthy volunteers) with a VRE phage administered intravenously. That was further than his rivals had gotten. He expected his phage to be in phase two trials soon. In animal models, he added, the phage had proven vastly superior to Synercid and somewhat superior to Zyvox. His new goal was to use phage *in conjunction* with these and other antibiotics: each therapy might help the other.

As they raced to market, all three start-ups had also explored how their old approach might be helped by the newest science in the healthcare field: genomics. What would happen when their phages were sequenced, when their DNA was spelled out as clearly as lettered beads on a string? Would genomics help them refine their phages and expand their range of efficacy? Would it give them new tools to keep phages from being sequestered in the liver and spleen? At the least, they hoped, it would prove to the FDA that their phages were harmless, and yield a blueprint they might use to win patents on modified phages from the U.S. Patent and Trademark office. Accordingly, all three start-ups were getting their phages sequenced.

For all three start-ups there was hope, but years would pass before any of them would get an FDA-approved phage to market. In the meantime, a window of vulnerability, as pharmaceutical executive George Miller put it, had opened, and would stay open, at least from 2002 to 2007: a time in which few if any drugs of the caliber of Zyvox or Synercid seemed likely to emerge, even as multidrug-resistant bugs grew tougher to beat.

Like peptides, phages offered a bright new hope. But neither was a magic bullet. Not yet, at least. Not soon enough. Not soon enough at all.

15

PEERING INTO THE ABYSS

In the winter of 2002, multidrug resistance was on the rise in all the usual bacterial suspects, both Gram-positive and Gram-negative, that plagued U.S. and European hospitals. One of the new threats, community MRSA, had spread from Minnesota to several other U.S. states. Half a dozen more cases of vancomycin-intermediate-resistant *S. aureus*, or VISA, had cropped up around the United States after the initial two cases in 1997, then seemed to peter out, though experts were divided on what the lull meant. Some thought VISA had proved an evolutionary "dead end" for *S. aureus*. Others, among them the CDC's Fred Tenover and Japan's Keiichi Hiramatsu, believed that MRSA strains were growing gradually less susceptible to vancomycin as the small "outlaw bands" of resistant cells within human isolates grew larger. "Eventually, we are going to hit the threshold," Tenover said. "I can feel it. And once that happens, it will spread like crazy." But to doctors who looked at infectious diseases on a global scale, the most worrisome news about multidrug resistance concerned a pathogen most Westerners dismissed as ancient history: tuberculosis, the airborne killer.

In recent times, TB had infected relatively few Americans, tamed as it was by antibiotics, and for those who did contract it, a phalanx of ten or more drugs stood ready to battle it. But this unique bacterium, neither a Gram positive nor a Gram negative but a *Mycobacterium* with its hard, polysaccharide capsule, had risen again to terrorize much of the world in a form more potent than ever before seen in history: "Ebola with wings," as some doctors put it. Strains were commonly resistant to all four of the first-line drugs used against it, as well as to each of the remaining second-line drugs. Most alarmingly, it had formed a dire alliance with the human immunodeficiency virus. Already second to AIDS among the world's most lethal infectious diseases, TB now killed as many as 2.8 million people each year. The crisis was at its most acute in Russia: the epicenter from which multidrug-resistant TB, or MDR TB, was all too likely to spread, traveling to the West like a rag-clad, foul-smelling, unwanted guest. It might not be an epidemic this year in the United States, perhaps not in five years—but very possibly in ten.

Historically, the Soviet Union had controlled TB with ruthless effectiveness. The Sanitation and Epidemiology Service, known as SanEp, forced all citizens to undergo chest x-rays for the disease as part of annual health checkups for all manner of infectious diseases. Anyone who tested positive for active TB was immediately removed from home and family, and quarantined in a government sanitorium, there to undergo chemotherapy for an average period of two years. As a result, TB rates in the USSR declined from the early 1950s, as they did in the West, even with the hardships and stress of life under Communism, pressures that might otherwise have exacerbated the spread of the disease.

Unfortunately, when the Soviet Union collapsed, so did SanEp. General health declined and poverty and crime soared, along with alcoholism, drug addiction, and prostitution. Of those ills, rampant and unprotected sex was the most dangerous. HIV spread. So did syphilis. The open sores of syphilis promoted the spread of HIV, which ravaged immune systems and reactivated latent TB. From TB came MDR TB.

In post-Communist Russia, prisons swelled with the arrested, many awaiting trials for as long as eighteen months, more than enough time in which to contract TB—in effect, a death sentence before judgment. Each year, 300,000 new prisoners were incarcerated. The national mortality rate for TB in Russian prisons was 484 per 100,000, among the highest in the world. Just as chilling, some 300,000 prisoners were let out each year into the community. Of those, roughly 30,000 had full-blown TB, including perhaps 7,500 who had MDR TB. In one year a single active TB carrier in the community would infect roughly ten other people. In all, one in every thousand Russians was now estimated to be actively infected with TB—fifteen times the U.S. rate—and 25 percent of all TB cases in Russian prisoners were multidrug resistant.

Paul Farmer, a forty-one-year-old Harvard-trained doctor, had seen more multidrug-resistant TB than most Americans, first at the clinic he maintained in Haiti through his Harvard-based group Partners in Health, then in Peru at another PIH clinic, where epidemic strains proliferated, and now as an adviser to currency billionaire George Soros in the latter's effort to contain TB in Russian prisons. Farmer had played an important role in helping to convince Soros—as well as the World Health Organization—that treating Russian prisoners with the standard first-line drugs for simple TB and what was called DOTS, Directly Observed Therapy, Short-Course, would not only fail to cure them but actually worsen the crisis. DOTS was based on the recognition that patients, no matter how well intentioned, would fail to take their drugs on a regular basis and so squander the chance to get well; with DOTS, patients would be forced to take their drugs in a healthcare worker's presence. The DOTS protocol worked well for simple TB: the four first-line drugs would be given for two months, followed by two other drugs for four months.[1] But an MDR TB patient was resistant—by clinical definition—to at least two of those drugs. Almost inevitably, the MDR TB proved wily enough to grow resistant to the others as well. The more drugs the bug resisted, the more it proliferated. Accordingly, DOTS tended to exacerbate rather than contain MDR TB. Farmer championed what

[1]Usually the four drugs first used were rifampin, isoniazid, ethambutol, and pyrazinamide.

he called DOTS PLUS. Each MDR TB patient was tested for his or her particular pattern of susceptibility to the eleven first- and second-line drugs, then treated with the ones that worked against the strain. Because MDR TB was so much more virulent than simple TB, treatment took up to two years, all under the watchful gaze of a healthcare worker, as per DOTS. Even in Russia, the cost of treating MDR TB in prisons was high: about five thousand dollars a case versus a few hundred dollars for simple TB. But Farmer had convinced Soros that DOTS PLUS was essential.

Already, strains of MDR TB had been borne across Russian borders into eastern Europe. Anywhere else was only a plane ride away. One day in the fall of 1998, a Ukrainian émigré flew from Paris to New York, coughing repeatedly en route. No flight attendant took note; no U.S. customs agent stopped him from passing into the country. Two days later, the man checked himself into a western Pennsylvania health clinic. A chest x-ray and culture test showed he had active tuberculosis; cultures eventually revealed TB resistant to six drugs. When health investigators tracked down the forty passengers who had occupied seats in the man's vicinity, what they found appalled them. Thirteen other passengers had positive tuberculin tests. Eventually, lab tests would show they had all been infected by the Ukrainian passenger.

Occasionally at one of the U.S. medical conventions, a drug company would report progress on a prospective new antibiotic to combat the creeping crisis of global MDR TB. But the truth was that *no* new drug lay just over the horizon. As far as Paul Farmer could tell, no new *kind* of drug was in the pipeline at all. The few distant prospects were variations on existing drugs.

Why had no new TB drugs come along in more than thirty years? Because, as Farmer knew all too well, TB was viewed as a poor man's disease and big pharmaceutical companies had no wish to spend half a billion dollars developing a drug that had no lucrative market. Worse, Farmer claimed, more than one drug company's executives all but pleaded with him not even to try their new wide-spectrum antibiotic on TB. If it worked, the company would be pressured by international agencies—and mavericks like Farmer himself—to distribute the drugs to the poor at wholesale cost or no charge.

Big pharma *had* produced two new drugs for multidrug-resistant *S. aureus* and other Gram positives. But in their nearly two years on the market, Synercid and Zyvox (the commercial name for linezolid) had had mixed results. At almost twice the cost of its rival, Synercid sold poorly; some doctors already referred to Synercid in the past tense, claiming its efficacy was modest at best. Because the less expensive Zyvox had an oral as well as intravenous form, one new study suggested it was actually lowering hospital costs for MRSA: patients could check out sooner and take oral Zyvox at home. But with wider use, Zyvox had engendered the specter of significant resistance. In July 2001, an eighty-five-year-old man in London became the first person to die of Zyvox-resistant MRSA; the patient had been intolerant of vancomycin, and when Zyvox failed a number of other drugs, including Synercid, also failed. In December 2001, researchers at Minnesota's Mayo Clinic reported a case of Zyvox-resistant VRE. More worrisome, the resistant strain transferred to six other patients, presumably by contact. The Mayo doctors recommended that in other such cases, a brand-new drug called daptomycin might work.

Daptomycin was indeed the bright new hope for multidrug-resistant infections—the presumed successor to Synercid and Zyvox. It had an interesting story. Based on the extract of a soil fungus discovered and developed by the Eli Lilly company, it had seemed promising but then showed signs of toxicity that led Lilly to put it on the shelf. Some years later, a Lilly researcher sought a job transfer to a tiny start-up company called Cubist in Cambridge, Massachusetts, and described the drug in his interview. Cubist had gone public on the promise of finding new protein targets for large pharmaceuticals, only to run into "picket fences" of patents put up by SmithKline Beecham. Now the start-up was desperate for a drug it could get to market before its backers grew restless. Cubist's Frank Talley, who interviewed the job applicant, immediately understood the politics that played into the story: an internal group of executives at Lilly wanted a successor to vancomycin with a vancomycin-sized market, and they seized on the first signs of toxicity in daptomycin as an excuse to shelve the drug because its market potential seemed more modest. (Ironically, the son of vancomycin they chose to pursue fizzled in development.) At Talley's urging, Cubist bought the unwanted-stepchild drug, tested it at lower doses, at once a day rather than twice, and found it was a winner after all. "It's the most cidal drug anyone's ever seen against MRSA and VRE," Talley declared three years later. Yet in January 2002, the company reluctantly announced that Cidecin, its brand name for daptomycin, had failed its phase three trials for community-acquired pneumonia, throwing the drug's future into doubt and Cubist's stock into a tailspin.

In medical meetings of late 2001, other new drugs in the pipeline were touted: a new "oxy" that might challenge Zyvox, new fluoro-quinolones, and more. But the truth was that aside from Cidecin, no new broad-spectrum antibiotic was close to market. In the past three years, several major pharmaceutical companies had actually bailed out of the business of developing new antibiotics. Eli Lilly was out, apparently discouraged after its son of vancomycin failure. So were Roche and DuPont, while Bristol-Myers Squibb and GlaxoSmithKline were both said to be considering termination of their antibiotic R&D programs. The bottom-line choice, even without factoring in the inevitability of resistance, was all too clear: return on investment from producing an antibiotic that might be used by a patient for less than a week versus return from a drug for a chronic condition that a patient might take daily for fifty years. That fall, a group of researchers from Tufts concluded that the average cost of developing a new drug had more than doubled since 1987—to $802 million. Consumer groups muttered that the study was biased and that the figures were inflated to justify high retail drug prices. But even slicing $100 million or more off the top, the challenge of rallying a company to invest in new antibiotics seemed formidable. Some observers, like Pharmacia's Chuck Ford, thought the only solution was for the government to extend the number of years a company could hold the patent on an antibiotic to increase the financial incentive for research and development. But politically, the prospect of federal handouts to the industry that already charged consumers staggeringly high prices for drugs seemed unlikely at best. Meanwhile, resistance was pushing up the cost of treatment with existing drugs to an alarming amount. A single course of antibiotics to treat a susceptible urinary tract infection cost about twenty dollars. To treat a UTI that was multidrug resistant required newer, more expensive drugs that cost about a hundred dollars a course. Despite the increase, drugs once used as last resorts—like vancomycin—were increasingly being used as first choices, with predictable results.

Where *would* other new drugs come from? The great glittering prospect was genomics. In the last half of the 1990s, it had become a mantra, like *plastics* in the movie *The Graduate*. No longer was it just a scientific term that meant theorizing about how the genes of a cell might fit together. New, computer-driven sequencing machines analyzed entire genomes of living things: viruses, bacteria, animals, and humans. (Corn, rice, tomatoes, and sweet peppers, too.) With that new technology, scientists could start seeing, so much more quickly and clearly, which genes did what, how they worked together to create an exact duplicate of the cell, and—of greatest interest to the pharmaceutical industry—how they caused disease or allowed it to occur.

J. Craig Venter, founder of The Institute of Genomic Research, had kicked off the gold rush in 1995 by publishing the first entire sequence of a genome—the bacterium *Haemophilus influenza.* Since then, an entire industry had materialized. Much of it was inspired by the race to sequence the human genome—a race in which Venter was pitted against the federal Human Genome Project, led by Nobel Prize–winning scientist James Watson. But even before that race ended in the spring of 2000, with both sides claiming victory, TIGR and a number of biotech start-ups had used a new method of Venter's called shotgun sequencing to analyze the genomes of dozens of other bacterial pathogens—each, like *Haemophilus,* containing about 2,000 genes as opposed to the generally estimated 40,000 genes in the human genome. These smaller genomes, free of the "junk" DNA that took up much of the human genome, were incredibly promising fields of study. All a drug company had to do was determine which of those genes was important to the bacterium, and—presto—it would have new targets for antibiotics. Or so went the hype.

By the turn of the century, every major drug company in the world was deeply invested in genomics. Using high-speed computers, researchers scanned their newly sequenced pathogens to see which genes appeared to play vital roles in virulence, resistance, and replication. Using new, computer-driven tools like high-through-put screening and combinatorial chemistry, they created vast libraries of chemical compounds, every one a bit different from the others, to try against all those targets. Soon, instead of not having enough new drug targets, the researchers were *inundated* with them.

With that, reality sunk in. The drug companies realized they could hardly pursue all of these potential targets, or even more than a few, through the years-long, vastly expensive gamut of drug development. So they started sifting through promising genes to find the most nearly perfect ones. For starters, they considered only entirely new *kinds* of targets—ones that would lead to entirely new classes of drugs. But with such an embarrassment of possibilities, being novel was hardly enough. The best candidates ought to appear in various kinds of bacteria—though only in bacteria, of course, not in man—so that a compound used against them would have a broad range of action. The researchers also looked to be sure a target gene was expressed in vivo. No point in targeting a gene that lay dormant while the bug was infecting its host. Finally, the target had to be essential to the bug's survival.

By the winter of 2002, the researchers had "knocked out" promising genes from *S. aureus* and a host of other pathogens. Still, a pall had settled over the newborn industry. Those targets in hand were promising, but their functions remained vague. "Everyone is talking about being so target-rich," scoffed one senior drug company executive not involved with genomics. "Nobody knows what they do!"

Even when a perfect target was determined and the perfect chemical compound was found to zap it, a drug company would still need to spend many additional years putting the drug through three-phase clinical testing, because, as George Poste, former chief scientist of

SmithKline Beecham, put it, genomics solved only the front-end problem of drug development. Indeed, with the several years that genomics added to the front end, a new drug might take even *longer* to reach the market than it had in the pre-genomics era. And in clinical trials, any drug, even one determined by genomics, might cause an unexpected adverse reaction in a few clinical volunteers. All it took was a few such adverse reactions to kill the work, and investment, of years. No one doubted that one day genomics *would* yield a new generation of antibiotics. But that day was, by the most optimistic estimates, half a dozen years away.

François Bompart, manager of one phase of Synercid's development, felt he knew a faster route than genomics to heading off resistance. He had transferred from antibiotic development at Aventis—successor to Rhone-Poulenc Rorer after a megamerger—to focus on vaccines. Like Paul Farmer, Bompart was driven by a strong humanitarian streak. "The world of Synercid," he says, "is that of high-tech medicine, dealing mainly with ICU patients, expensive and performed only in very few places in the richest countries of the industrialized world. Vaccines oblige you to consider issues on a global scale."

Bompart was studying the process by which *S. aureus,* when it invades the body, creates a protein known as RAP. Once a certain threshold level of RAP has been produced, the staph bacteria begin producing their deadly toxins. A RAP vaccine, Bompart reasoned, might train the immune system to neutralize staph's RAP and so disarm the bacteria. Work was also under way to develop a vaccine for TB, thanks in large part to a $25 million grant for vaccine development from the Bill and Melinda Gates Foundation, along with a $25 million grant for TB study. The other boost to the field was the phenomenal success of Prevnar, the conjugate vaccine for *S. pneumo* suitable for young children. In its first year, Prevnar had generated sales of $461 million for Merck; for doctors, it was a way to avoid having to use antibiotics against which *S. pneumo* was growing increasingly resistant. Still, Prevnar's cost was sobering: $232 for a four-dose round. The average childhood vaccine cost $10.

Vaccines were a mainstream alternative, phages a radical long shot. In between were other alternatives gaining cachet as resistance rates rose. Stuart Levy had helped found a company called Paratek to develop the equivalent of "smart bombs" that would attack microbial mechanisms of resistance. Levy's main target was efflux pumps. If he could disable them in various bacteria, then the drugs that they pumped out—tetracyclines chief among them—would become effective again. Levy's product would be administered in conjunction with those antibiotics.

As a spokesman for prudent antibiotic use, Levy argued that doctors reached too often for broad-spectrum drugs. For empiric therapy those were the safe bet, ostensibly wiping out whatever bug might be causing a patient's infection. But broad-spectrum drugs also eradicated harmless or even beneficial bacteria. Faster and more accurate diagnostics would alleviate this problem. At New York's Public Health Research Institute, Barry Kreiswirth and colleagues were developing tests to identify almost instantly the exact bacterial strain infecting a patient. The tests would also reveal a bug's profile of drug resistance. Doctors could use narrow-spectrum antibiotics with confidence, curing a patient without damaging healthy flora, as well as saving broad-spectrum drugs for when they were most needed.

None of these alternatives was market-ready. But at the CDC, Theresa Smith's boss, Bill Jarvis, head of the agency's infectious diseases outbreak group, felt sure he had a way to curb the spread of drug-resistant infections today—without three-phase trials, without a half-billion-dollar investment, without even a product. Infection control was a lot less exciting than a brand-new drug, and most doctors privately felt that no amount of hand washing would prevent the spread of hospital bugs like VRE, but Jarvis was a contrarian—and now he had proof.

In February 1997, the first few cases of VRE had appeared in hospitals of the Sioux Lands—a fuzzily defined region spanning parts of Nebraska, South Dakota, and Iowa. When Steve Quirk, a public health director for part of the region, called the CDC for help, Jarvis realized he could test a theory he'd been nursing ever since Glenn

Morris's failed effort to contain VRE at the Baltimore VA medical center. Morris had tried to contain VRE in only two wards of his hospital; inexorably, new VRE cases kept coming into the hospital from other institutions. In the Sioux Lands, Jarvis saw a chance to try infection control on a much larger scale: by having the region's thirty-two hospitals and nursing homes reveal exactly which of their patients had VRE and isolate all those who had it, so that the bug's spread might really be contained.

For that, the hospitals had to be candid—not an easy choice—and trust that Quirk, point man for the project, could keep confidential the patient information they handed over; none wanted newspaper articles declaring that one institution had more VRE than another. To help, Quirk assigned each institution a code number based on one of thirty-two cards that his two-year-old son flipped up in the air. All VRE cases were identified and isolated, as were all VRE patients transferring from one institution to another; new patients were also screened. Over the three years that the experiment was conducted, VRE in the region declined from 2.5 percent of all patients to .5 percent.

In fact, the Sioux Lands hospitals merely followed CDC guidelines. So few others were even trying to observe them, however, that Jarvis publicly lamented, at a medical convention in Atlanta, that perhaps the guidelines should be put aside. "Hey, what if John Paul Jones had had that attitude," a doctor in the audience responded, only half-joking, in a thick Mississippi accent. "Where would we be now?"

The speaker was Barry Farr, an infectious disease specialist at the state university hospital in Charlottesville, Virginia, who looked, and talked, a lot like media mogul Ted Turner. As much a maverick as Turner, he had made a mission of culturing all incoming patients—rare at any U.S. hospital—and isolating not just the patients infected with VRE or MRSA, but those *colonized* with the bugs. By doing so, he had cut infection rates so sharply that he'd entered into talks with forty of the hospitals nearest his own to try to extend his policies, just as Jarvis had done in the Sioux Lands.

Before long, the two men were scheming to launch a two-*state* program covering Virginia and North Carolina, a catchment area of some 300 hospitals and 650 nursing homes. By late 2001, the program was up and running. Convinced now that failures in infection control were actually *more* responsible for spreading resistant bugs than overuse of antibiotics, Jarvis and Farr felt sure they could slash VRE and MRSA rates in both states. When they did, perhaps other hospitals would no longer be able to ignore CDC guidelines for infection control. The guidelines might acquire the force of regulation; they might even become federal law. But even within the CDC, the program had its skeptics—those who believed that overuse of antibiotics, not infection control lapses, was the principal cause of the problem. "We're fighting against a tide of conservative opinion on this," Farr admitted. But like John Paul Jones, he felt he had just begun to fight.

Another easy way to curb resistance in the hospital might be cycling. Bugs grew resistant to drugs with repeated exposure to them, so why not switch the drugs periodically and throw the bugs off? The oft-cited case was Communist Hungary. By the early 1980s, 50 percent of the country's *S. pneumo* cases had proved resistant to penicillin. From 1983 to 1992, doctors switched to other classes of antibiotics; penicillin use plummeted and so did *S. pneumo* resistance to the drug, down to 34 percent. In the United States, cycling programs had met with mixed results, but Fred Tenover felt the studies merely suffered from poor controls: although a beta-lactam drug might be temporarily shelved at a particular hospital, patients continued to be exposed to penicillinlike compounds in the food supply or in water runoff from orchards sprayed with the drug. Immunologist Marc Lappe agreed that cycling would work—but only if carried out across the board. "Antibiotic resistance would dissipate to a substantial degree," he declared, "if you excluded certain antibiotics from the marketplace." But Bruce Levin, a prominent population biologist at Emory University in Atlanta, Georgia, felt that with many bugs and drugs resistance was, as he put it, a one-way street. His own studies suggested that once resistance is established, the strains that acquired it tend to hold on to it. If it waned at all, he felt, it did so modestly— a shocking conclusion. And if antibiotic use resumed, resistance in-

creased sharply. "We are committed to an arms race," he said of hu-
mankind's battle with resistant bugs. Disarmament, he added, was
not an option.

Of all the possible ways to curb the spread of drug-resistant bacteria,
surely the easiest, and most compelling, was to ban growth promoters
in agriculture. In October 2001, Fred Angulo and Henrik Wegener
both contributed to a series of reports in the *New England Journal of
Medicine*. As an attached editorial observed, the new studies represented "the proverbial smoking gun" that growth promoters produced animal-to-human transfer of resistance.

Angulo and colleagues tested chicken carcasses across the United
States in 1998–99 and found that 58 percent of the 400 chickens they
tested had drug-resistant *E. faecium*. Worse, 17 percent of them contained Synercid-resistant strains of *E. faecium*. Of course, no chickens
had been fed Synercid; the growth promoter virginiamycin had provoked cross-resistance to its fellow streptogramin. Angulo's team also
found Synercid resistance in 1 percent of the human stools they
cultured—a damning show of animal-to-human transfer, given that
few if any of those people had been treated with Synercid, and one
that only confirmed Marcus Zervos's initial studies.

Wegener and other researchers at Denmark's Staten Serum
Institute performed a brilliantly simple test to determine just how
long resistant animal enterococci stayed in the human gut. Twelve
healthy volunteers whose stools showed no evidence of drug-resistant
enterococci ate four different strains of resistant enterococci derived
from recently slaughtered chickens or swine. Six other healthy volunteers ate animal enterococci that were drug susceptible. For five days
afterward, the resistant microbes showed up in the stools of all dozen
of the first group, long enough to interact with human enterococci
and pass along their resistance genes.

In the United States, Stephen Sundlof awaited the results of the
CVM's study of virginiamycin but seemed disinclined to act on any
existing growth promoters. Very possibly, the growing public clamor
against them might curb their use without his intervention. In

February 2002, a front-page story in the *New York Times* announced that three of the country's largest poultry producers—Tyson Foods, Perdue Farms, and Foster Farms—had reduced or eliminated the use of growth promoters quietly over the last several years. But there was not yet any indication on poultry packages to inform consumers if chicken products from these or other companies were actually free of growth promoters, and patrons of fast-food joints or restaurants could not be sure the chicken they ordered was growth-promoter free, either. It was a good step, but only that.

As for Sundlof's proposed ban on quinolones for therapeutic veterinary use, it remained just that, a proposal, given that Bayer was still contesting it. The company that manufactured Cipro—the drug had been its biggest seller in 2000, with profits of about $1.6 billion—was lobbying hard to keep selling an almost identical drug for animals that had jacked up quinolone resistance in poultry from 1 percent to more than 19 percent in the four years it had been used. Continued use of Baytril would certainly shorten the lifetime of Cipro as quinolone resistance in animals spread and transferred to humans. But Bayer, reeling after a series of market setbacks and thought to be an acquisition target by even bigger drug companies, apparently wanted the short-term gain of selling Baytril, even at the risk of greater long-term costs.

The most profound way to combat antibiotic resistance might also be the hardest. For nearly a century, Western medicine had done its best to obliterate infectious diseases. To a remarkable degree, it had succeeded, at least temporarily. For years, the antibiotic era had seemed the apotheosis of that crusade: elimination of every bacterial pathogen from the face of the earth. The rise of multidrug resistance in the 1990s had forced science to see that no antibiotic would be a magic bullet for long. As long as doctors relied on such specific mechanisms, those one-celled creatures would always find equally specific countermechanisms to resist them. In a sense, antibiotics were our form of vertical resistance, which the bugs just met in kind. Yet science responded by getting even *more* specific: using genomics to find new antibiotic targets on a molecular level. Richard

Levins, a population biologist at Harvard, had preached for some time the need to adapt broader and subtler means of resistance—*horizontal* means. It was part of a different, more benign, more evolutionary view of bacteria. And it made an awful lot of sense.

Levins began with a question that Western science never considered. What did the bugs want? Answer: to get a good meal, avoid getting killed, and have an exit strategy to reach the next human host, preferably reproducing as they did. These three goals were often in conflict with one another, so bugs had to juggle. The bloodstream, for example, was the best place to get a meal, rich with nutrients and proteins. But it was also the most dangerous place in the body for a bug to be, because there macrophages and other elements of the immune system stood ready to kill it. So a bug might escape to the central nervous system: good meal, relatively safe, but no exit. Failing that, a bug might migrate to the skin. Now it had an easy exit to the next host and was relatively safe, given the scant circulation of the immune system to the skin. But the grazing was poor.

Given that none of these options was perfect, a bug tried to change the odds. It might try to make the blood safe, either by destroying the host's immune system or by becoming invisible to it. As it began to wreak havoc, a bug had to decide whether to reproduce there or leave first and reproduce later. The milder the symptoms it provoked, the longer a bug might be encouraged to stay—reproducing, but doing little harm. If strong, broad-spectrum antibiotics arrived on the scene to zap a whole field of pathogens, the toughest would do the best: selective pressure. But what if bacterial infections with mild symptoms were treated symptomatically, making the patient comfortable without antibiotics? If antibiotics were saved for the most severe cases, Levins theorized, then natural selection would favor the pathogens that produced the milder symptoms. This evolutionary strategy, in turn, would influence a pathogen to be less resistant.

Most microbes, Levins believed, were not inherently pathogenic, any more than they were inherently resistant to drugs. They merely wanted to survive. Along with drastically reducing the onslaught of antibiotics that forced them to grow tougher and fiercer, Levins preached the need for societal measures having nothing to do with drugs. A fundamental principle of ecology, he observed, is that organisms that are stressed become more vulnerable. The poor, the marginalized, the targets of racism, all have their homeostasis—the capacity of the body to detoxify—undermined. Our built environment can be a factor, too, Levins felt. The canned air of high-rise buildings, the increase in rat populations as garbage collection goes down, chemical pollution for people living near incinerators—all these can undermine the body's immune system and make it more vulnerable to pathogens. Instead of reducing the problem of antibiotic resistance to the search for new molecules that might make the best next drug targets, Levins encouraged exploration of various means of broad resistance. None might confer complete resistance on its own, but collectively such measures might build a threshold of health over which bacterial pathogens would have a harder time clamoring.

An evolutionary approach, Levins felt, might keep the antibiotic era from ending. An unrelenting focus on new magic bullets would only hasten its close. Why should *S. aureus,* he observed, be content with intermediate resistance to vancomycin if, amid heavy and growing use of the drug, full-blown resistance suited it so much better? Ecologically, why should it not succeed in acquiring those resistance genes from VRE, perhaps through an intermediary like the ubiquitous *E. coli?* Throughout history, Levins observed, successful organisms have made toxic substances tolerable, even necessary: over time, for example, the early antibiotic streptomycin had become a nutrient to various bacteria. "Unless we play the evolutionary game consciously, we're going to be outwitted," Levins warned. "The brute force of designing new antibiotics won't keep us ahead."

Most of the time in his Baltimore lab, Glenn Morris tried to be more optimistic than that: he felt he had no choice but to be. He hoped the phages he was working on would save the day, or at least give doctors a new choice. He hoped that by cutting back on antibiotics at last, doctors might stop rates of resistance from rising, or perhaps even get them to dip. He saw real hope in the gathering public outcry against growth promoters and the possibility that they might be banned altogether. But Morris was a realist. He knew that doctors would likely keep throwing tens of millions of pounds of antibiotics a year at bacterial infections. He knew that resistance genes, once formed by mutation and spread by selective pressure, would probably linger out there in the miasmic netherworld of one-celled pathogens—dropped perhaps by some bugs if antibiotic use dipped, but picked up by others. A lot of damage had been done these last sixty years, a lot of damage that could possibly not be reversed.

Morris was a scientist, and a doctor, and he stuck to the facts. But sometimes, at the end of a long day in the lab, looking through a microscope at bugs that spent every second of their lives trying to protect themselves and replicate, Morris got a creepy feeling. On some instinctive evolutionary level, *the bugs seemed to have figured it out.* No matter what we tried to kill them with, they almost seemed aware now that they could fight it off. Worse, given how fast rates of resistance were rising, it was as if they'd spread the word among their trillions of brethren: *we can beat this new threat called humans.*

Morris always felt relieved, when he had those premonitions, to turn out the light in his laboratory, go down to his car, and be a commuter in the here and now. But he never quite shook the suspicion that if we kept on trying to obliterate drug-resistant microbes in the all-out, reckless way we had for the last sixty years, there would be only one outcome.

The microbes would win.

Just three months later, Morris scanned the top headline of the CDC's *Morbidity and Mortality Weekly* of July 5, 2002, with shock and alarm. The microbes had staged an appalling breakthrough. It was, indeed, the news that doctors had been dreading for a decade. In sober medical terms, the CDC reported that a forty-year-old Michigan man with diabetes, vascular disease, and chronic renal failure had just tested positive for full-blown vancomycin-resistant *S. aureus,* or VRSA. This wasn't a case of the bug creating a more swollen cell wall. This wasn't about

vancomycin-*intermediate*-resistant *S. aureus,* or VISA, getting incre-
mentally worse. This man's *S. aureus* had done the unthinkable: it had
acquired the *vanA* resistance gene from enterococci. Not only that: it
had acquired a *mecA* gene of resistance to oxacillin as well.

True, Morris read on, the man's infection remained susceptible to
Zyvox and Synercid, along with a few other antibiotics. But the essential transfer had occurred. The gene that Patrice Courvalin had first
identified in enterococci, the one that William Noble had coaxed into
transferring from enterococci to *S. aureus* in the lab, had made the leap
in vivo at last. It had done so, Morris read, at an infected catheter site.
The cultures were astounding. By comparison, in the eight documented U.S. cases of VISA, the minimum inhibitory concentration of
vancomycin had been 8 ug/mL. The MIC for oxacillin with this new
bug was greater than 16 ug/mL. The MIC for vancomycin with it was
greater than 128 ug/mL.

Using "aggressive wound care" and a regimen of trimethroprim-
sulfamethoxazole, the Michigan man's doctors had healed the
catheter site infection. Since then, it had tested negative for VRSA. To
date, the CDC reported, no evidence of transfer had occurred among
healthcare workers or other patients. So this was one case—a case of
random bad luck. Perhaps it would never recur.

But that, as Morris well knew, was not the way bacteria worked.
Now that the trick had been accomplished, it would happen again—
and again. In a year, a few more cases would be reported. A year or
two after that, the rates would start to rise by a significant percentage,
though still seem low by an absolute measure. And then—in three
years? in five?—would come the breathtaking surge, held back only
in part by the two new stopgap drugs, Zyvox and Synercid, against
which resistance had already appeared.

Glenn Morris shook his head. Bit by bit, bug by bug, things were
slipping away. And the bacteria were everywhere. Back in 1995,
Morris had thought that things couldn't get worse. But they had. And
barring that miracle discovery, there was no reason to believe that
things would get better anytime soon.

Glenn Morris would not sleep well that night.

ACKNOWLEDGMENTS

To embrace a field as broad and complex as drug-resistant bacteria, with all the microbiology it entails, then try to make it simple, clear, and compelling, we needed a lot of wise counsel. J. Glenn Morris Jr. of the University of Maryland's Baltimore Veterans Affairs Medical Center was one such guru; we are grateful not only for his considerable help but for his commitment to fighting the war against drug-resistant bacteria on so many fronts. Like Morris, Fred Tenover and Fred Angulo of the Centers for Disease Control both consented to several interviews, with much follow-up e-mailing, effortlessly ticking off the subtleties of human and animal antibiotic resistance. Knowing that two such capable experts are on the case for the U.S. government—committed public servants both—is one of the few cheering aspects of this scary situation. Keiichi Hiramatsu of Tokyo's Juntendo University is another world-class expert in the field who was kind enough to offer considerable help, both in person and in long follow-up e-mail messages. Barry Kreiswirth of the New York Public Health Research Institute spent many hours with us; so did Chuck Ford, senior scientist at Pharmacia & Upjohn. Henrik Wegener of the Danish Zoonosis Center was a gracious host when we visited him in Copenhagen; he then kindly read and commented on our chapters on resistance in animal-use antibiotics. Christina Greko, in Stockholm, was just as gracious and helpful. Nina Chanishvili spoke with us at length when we visited her at the Eliava Institute in Tbilisi, Georgia; she then responded to our follow-up queries with many long, very helpful e-mail messages.

Several other experts kept up a regular e-mail correspondence with us, alerting us to new studies, answering our queries, and then reading much or all of our manuscript. Jon Blum of the Harvard Medical School was extraordinarily helpful in this regard, especially in guiding us through the thickets of Patrice Courvalin's work with vancomycin-resistant enterococci, which made up the most scientifically complex chapter of our book. Yale Medical School's Christopher Herndon was just as valuable a guru—a real fount of scientific information, and never too busy, it seemed, to respond to our queries. Susan Donelan, Director of Infection Control at Stony Brook Hospital in Long Island, New York, gave us an insider's tour, answered many follow-up queries, and read much of our manuscript with an eagle eye while somehow tending to newborn twins at the same time. Harriette Nadler of Rhone-Poulenc Rorer, Richard Goering of Creighton University, Tamar Barlam of the Center for Science in the Public Interest, and Steve Projan of Wyeth-Ayerst also answered many queries and offered much help along the way. Early on, we were buoyed by a grant from Byron Waksman and the Foundation for Microbiology.

In addition, we are indebted to the following for their considerable kindness and time in granting us interviews: Zamphira Alavidze, Evangeline Ames-Murray, Peter Applebaum, Thomas Aragon, Jeff Band, Kiran Belani, Barry Bloom, François Bompart, Lori Boschetto, Daniel Bouanchaud, Steven Brooks, Gary Burke, Colin Campbell, Richard Carlton, Richard Carnevale, Henry Chambers, Teimuraz Chanishvili, David Cole, Patrice Courvalin, Robert Daum, Gary Doern, Michael Dowzicky, Paul Ewald, Paul Farmer, Barry Farr, Celine Feger, Dan Feiken, Vince Fischetti, Karen Florini, Claire Fraser, Terry Fredeking, Hank Fuchs, Alfred Gertler, Steven Gill, Don Gillespie, Caisey Harlingten, William Haseltine, David Heymann, Richard Honour, Michael Jacobs, George Jacoby, Kathleen Jakob, William Jarvis, Judith Johnson, Wesley Kloos, Marin Kollef, David Landman, Marc Lappe, Robert Lehrer, Herminia de Lencastre, Bruce Levin, Richard Levins, Stuart Levy, Patty Lieberman, Don Low, Guido Majno, Nina Marano, Damien McDevitt, Don McGraw,

Antone Medeiros, Amiran Meipariani, John Mekalanos, Carl Merril, George Miller, Linda Miller, Carol Moberg, Robert Moellering, Tim Naimi, William Noble, Gary Noel, Tim Nolan, Richard Novick, Michael Osterholm, Lance Peterson, Sandy Pine, George Poste, Steve Quirk, James Rahal, Eric Reines, David Remy, Teri Remy, Peter Reynolds, Louis Rice, Tim Rieser, Barbara Robinson-Dunn, Mary Claire Roghmann, Hugh Rosen, Marty Rosenberg, Dan Sahm, Patrick Schlievert, Dean Shinabarger, David Shrayer, Caroline Smith de Waal, Kirk Smith, Theresa Smith, Lord Soulsby of Swaffham Prior, Steven Spiritas, Eric Spitzer, Dennis Stevens, Neal Steigbigel, Karen Stronsky, Alexander Sulakvelidze, Stephen Sundlof, Frank Talley, Michael Taylor, Judy Teppler, John Threlfall, Mona Tice, Mark Todd, Alexander Tomasz, Bill Trick, Carl Urban, Craig Venter, Chris Walsh, Patrick Warren, Rob Williams, Rosamund Williams, Mary Wilson, Wolfgang Witte, Rich Wood, Richard Wrangham, Michael Zasloff, Marcus Zervos, German Zuluaga, and Gary Zurenko.

At Little, Brown, Geoff Shandler adroitly edited the final manuscript, paring it and reshaping it to make it immeasurably better, and we are very grateful for that. Thanks also to Sarah Crichton, who was the first editor to believe in this book. Thanks to our agents, Joni Evans and Rafe Sagalyn, for their help throughout. Thanks to Liliana Madrigal for her love and support, and to our friends for enduring an awful lot of talk about vancomycin-intermediate-resistant *Staphylococcus aureus* and many other such microbiological mouthfuls.

NOTES

1: THE SILENT WAR

Facts about Glenn Morris and his clinical rounds at Baltimore's VA medical center are drawn from author interviews with Morris.

11 Every year, 1.2 million children around the world were estimated to die of *S. pneumo:* Remarks by Anthony S. Fauci, M.D., director of the National Institute of Allergy and Infectious Diseases, National Institutes of Health, before the Senate Committee on Health, Education, Labor and Pensions Subcommittee on Public Health and Safety, February 25, 1999.

12 Now 45 percent of all *S. pneumo* strains were penicillin resistant: "Antimicrobial Resistance with *Streptococcus pneumoniae* in the United States, 1997–98," by Gary V. Doern et al., *Emerging Infectious Diseases* 5, no. 6 (1999).

12 Bryan Alexander, eighteen, was found guilty: *Arlington* (TX) *Morning News,* February 22, 2001.

12 A few months later, talk show host Rosie O'Donnell: *Rosie* magazine, July 2001.

14 Stuart Levy, M.D., a Tufts University professor: "The Hunt Is On," by Mary Knudson, *Technology Review,* 100 (January 11, 1998).

14 Joshua Lederberg, M.D.: "Superbugs," by Sheryl Gay Stolberg, *New York Times,* August 2, 1998.

14 In 1954, 2 million pounds of antibiotics: "The Challenge of Antibiotic Resistance," by Stuart Levy, *Scientific American,* March 1998.

15 Yet researchers at the [CDC] judged that a full third: ibid.

15 Every year in U.S. medical institutions, 2 million patients: Centers for Disease Control, Atlanta, Georgia.

16 the estimated $5 billion cost of treating drug-resistant infections: Fauci, Senate comments, February 25, 1999.

16 high incidence of pathogenic bacteria on computer keyboards: "Computer Keyboards and Faucet Handles as Reservoirs of Nosocomial Pathogens in the Intensive Care Unit," Sergio Bures, M.D., et al., *American Journal of Infection Control* 28, no. 6 (December 2000).

16 Another had found the bugs in the cushions and fabric of chairs: "Persistent Contamination of Fabric-Covered Furniture by Vancomycin-Resistant enterococci," by Gary A. Noskin, M.D., et al., *American Journal of Infection Control* 28, no. 4 (August 2000).

16 A third had found them on rectal thermometers: "Hospital-Acquired Infection with Vancomycin-Resistant *Enterococcus faecium* Transmitted by Electronic Thermometers," by L. L. Livornese Jr. et al., *Annals of Internal Medicine* 118, no. 2 (January 15, 1993): 156; *Annals of Internal Medicine* 117, no. 2 (July 15, 1992), 112–16.

16 fourth on stethoscopes: "Contamination of Gowns, Gloves, and Stethoscopes with Vancomycin-Resistant Enterococci," by K. C. Zachary et al., *Infection Control & Hospital Epidemiology* 22, no. 9 (September 2001): 560–64.

16 In fact, one recent study conducted at Duke University: "Adverse Effects of Contact Isolation," by Kathryn B. Kirkland and Jill M. Weinstein, *Lancet* 354, no. 9185 (October 2, 1999).

18 "We can close the books on infectious diseases": *The Plague-Makers,* by Jeffrey A. Fisher, M.D. (New York: Simon & Schuster: 1994), 18.

19 Of those 50 million pounds of antibiotics: Levy, *Scientific American,* March 1998.

19 Each year, *Salmonella* infected: author interview with Fred Angulo of the Centers for Disease Control.

21 Treating a single case of multidrug-resistant tuberculosis: author interview with Paul Farmer of Partners in Health.

21 "We are seeing a global resurgence": U.S. Surgeon General David Satcher, in remarks to Congress, March 3, 1998.

2: IT'S A BUG'S WORLD

For the general facts about bacteria at the start of this chapter, we are indebted principally to I. Edward Alcamo's wonderfully readable textbook Fundamentals of

Microbiology. *Also very helpful was Stuart Levy's* The Antibiotic Paradox. *For general facts about* S. aureus, *we relied on those sources, as well as "The Genus* Staphylococcus," a chapter by Wesley E. Kloos, Karl-Heinz Schleifer, and Friedrich Gotz, from A. Balows, H. G. Truper, M. Dworkin, et al., eds., The Prokaryotes, *2d ed. (New York: Springer-Verlag, 1991); and the Food and Drug Administration's online "Bad Bug Book" entry on* S. aureus (http://vm.cfsan. fda.gov), *as well as the entry on same from the now-defunct Web site* www.ican prevent.com, *pioneered by former Minnesota state epidemiologist Michael Osterholm (a regrettable loss!). Also helpful were Jeffrey A. Fisher's* The Plague Makers *and Geoffrey Cannon's* Superbug.

Author interviews with the following also informed this chapter: Jon Blum, Ph.D., Richard Goering, Ph.D., Wesley E. Kloos, Ph.D., Marc Lappe, Stuart Levy, M.D., Guido Majno, M.D., Steve Projan, Ph.D., Patrick Schlievert, Ph.D., Fred Tenover, Ph.D.

31 In the Bible's book of Exodus: from Exodus 9:10.

31 Girolamo Fracastoro propounded it: *De contagione, Contagiosis morbus et eorum Curatione* (1546).

31 Ogston went on to show: author interview with Wesley E. Kloos.

32 Christian Gram, a Danish doctor, declared: Alcamo, *Fundamentals,* 75.

32 British scientist Alexander Fleming discovered: The Fleming story is told well in *Miracle Cure: The Story of Penicillin and the Golden Age of Antibiotics,* by Milton Wainwright, who discusses Ronald Hare's contribution to the lore. It's also discussed in *The Plague Makers* and *The Antibiotic Paradox.*

34 began by successfully re-creating Fleming's experiments: Florey and Chain's work is discussed in *Miracle Cure, The Antibiotic Paradox,* and *The Forgotten Plague,* by Frank Ryan, M.D. (Little, Brown, 1992).

35 "The greatest possibility of evil": *New York Times,* June 26, 1945, 21.

36 By 1946, 14 percent of the *S. aureus* strains: "The Genus *Staphylococcus,*" by Kloos et al., 1399.

37 Then in 1952 came a startling case in Japan: *Why Antibiotics Fail,* by Marc Lappe (North Atlantic Books, 1986), 74.

39 Ever since the discovery of streptomycin: The story of vancomycin is drawn from an unpublished Ph.D. thesis on the subject, "The Antibiotic Discovery Era (1940–1960): Vancomycin As an Example of the Era," by Donald J. McGraw (Oregon State University, December 1975). Some details also drawn from author's interview with McGraw.

3: EARLY WARNING

The first part of this chapter is drawn principally from author interviews with Glenn Morris; the second part, on avoparcin and VRE, is based on an author interview with Wolfgang Witte, Ph.D. Other sources interviewed for the chapter were Barry Kreiswirth, Ph.D., Barry Farr, M.D., Donald Low, M.D., Fred Angulo, Ph.D., and William Jarvis, M.D.

48 Dr. Don Low, chief of microbiology: Mark Witten, "Outbreak," *Saturday Night* magazine, May 1996.

48 Not long before . . . a patient at the Toronto Hospital: ibid.

4: THE GENETIC DETECTIVE

This chapter was drawn principally from author interviews with Patrice Courvalin, Ph.D., of the Pasteur Institute, Christopher Walsh, Ph.D., of Harvard University medical school, Peter Reynolds, Ph.D., of Cambridge University, and William Noble, Ph.D.

58 A grateful public underwrote it: François Jacob, "The Pasteur Institute," at the Nobel e-Museum Web Site, www.nobel.se/medicine/articles/jacob.

70 Noble's article, published in a scholarly journal: "Co-transfer of Vancomycin and Other Resistant Genes from *Enterococcus faecalis* NCTC 12201 to *Staphylococcus aureus*," by W. C. Noble, Z. Virani, and R. G. Cree, *FEMS Microbiol. Lett.* 93 (1992): 195–98.

5: NIGHTMARE COME TRUE

This first part of this chapter is based chiefly on author interviews with Keiichi Hiramatsu, M.D., Ph.D. The Michigan section is based on interviews with Fred Tenover, Ph.D., Theresa Smith, M.D., William Jarvis, M.D., Barbara Robinson-Dunn, Ph.D., Jeff Band, M.D., and Marcus Zervos, M.D. The New Jersey VISA story is based on interviews with Theresa Smith, Gary Burke, M.D., Rob Williams, M.D., Lori Boschetto, and Karen Stronsky.

77 Hiramatsu's report appeared, belatedly: "Characterization of Staphylococci with Reduced Susceptibilities to Vancomycin and other Glycopeptides," by F. C. Tenover et al., *Journal of Clinical Microbiology* 36, no. 4: 1020–27. See also "Dissemination in Japanese Hospitals of Strains of *Staphylococcus aureus* Heterogeneously Resistant to Vancomycin," by Keiichi Hiramatsu et al., *Lancet,* December 6, 1997, and "Contribution of a Thickened Cell Wall and Its Glutamine Nonamidated Component to the Vancomycin Resistance Expressed by *Staphylococcus* *aureus* Mu50," by Longzhu Cui et al., *Antimicrobial Agents and Chemotherapy* 44, no. 9 (September 2000): 2276–85. In addition, Hiramatsu has written chapters of academic texts on VISA, including "Vancomycin Resistance in Staphylococci," from *Drug Resistance Updates* (Harcourt Brace & Co., 1998), 135–50 and "Mechanisms of Methicillin and Vancomycin Resistance in *Staphylococcus aureus*," by Hiramatsu, Teruyo Ito, Ph.D., and Hideaki Hanaki, Ph.D., from *Balliere's Clinical Infectious Diseases* 5, no. 2 (July 1999): 221–42.

78 The call came one day in late July: author interview with Barbara Robinson-Dunn.

83 In her hotel room, Smith laid out all the charts: "Update: *Staphylococcus aureus* with Reduced Susceptibility to Vancomycin—United States, 1997," *MMWR* 46, no. 33 (August 22, 1997), 765–66.

Theresa Smith's more complete report on the Michigan cases appeared in the *New England Journal of Medicine* on February 18, 1999, as "Emergence of Vancomycin Resistance in *Staphylococcus aureus*" (340, no. 7, 493–501).

84 Campbell wanted her to call him on arriving: author interview with Colin Campbell, M.D.

6: TWO NOT-QUITE-MAGIC BULLETS

The story of Synercid is based chiefly on author interviews with Rhone-Poulenc Rorer's Daniel Bouanchaud, M.D., François Bompart, M.D., Michael Dowzicky, Ph.D., Celine Feger, and Harriette Nadler, Ph.D. The story of linezolid is based chiefly on author interviews with Pharmacia & Upjohn's Chuck Ford, Gary Zurenko, Mark Todd, and Dean Shinabarger.

91 On March 20, 1998: "*Staphylococcus aureus* with reduced susceptibility to vancomycin isolated from a patient with fatal bacteremia," by Sharon S. Rotun et al., *Emerging Infectious Diseases* 5, no. 1 (January–March 1999). See also "Superbugs," by Sheryl Gay Stolberg, *New York Times Magazine,* August 2, 1998.

92 Alexander Tomasz and Krzysztof Sieradzki demonstrated: "The Development of Vancomycin Resistance in a Patient with Methicillin-Resistant *Staphylococcus aureus* Infection," by Sieradzki et al., *New England Journal of Medicine* 340, no. 7 (February 18, 1999).

92 Sixteen-year-old Teresa Miltonberger's ordeal began: The *Seattle Times,* May 22, 1998, 1, has a good initial report on the tragedy. A medical update on Teresa Miltonberger appears in an Associated Press dispatch of December 18, 1998. A retrospective piece, pegged to the sentencing of the school shooter, appears in the *Oregonian,* November 11, 1999, 1.

97 Etienne and Bompart's team conducted phase one trials: Several details from this stage of the story are adopted from "Germ Warfare," by Dan Greenburg, *New York Magazine,* October 1993.

99 Synercid, Moellering conceded in his report: "In Vitro Activity of RP 59500, an Injectable Streptogramin Antibiotic, Against Vancomycin-Resistant Gram-Positive Organisms," by Linda A. Collins et al., *Antimicrobial Agents and Chemotherapy,* 37, no. 3, March 1993, 598–601.

100 The patient, a forty-six-year-old woman: "Quinupristin/Dalfopristin (RP 59500) Therapy for Vancomycin-Resistant *Enterococcus faecium* Aortic Graft Infection: Case Report," by Vivek S. Sahgal et al., *Microbial Drug Resistance* 1, no. 3 (1995): 245–47.

102 Some 2,000 patients in all were given Synercid: Associated Press report, September 21, 1999; also, *Pittsburgh Post-Gazette,* September 9, 1997, sec. A, p. 3.

102 Even Robert Moellering found reason: "The Efficacy and Safety of Quinupristin/Dalfopristin for the Treatment of Infections Caused by Vancomycin-Resistant *Enterococcus faecium,*" by R. C. Moellering et al., *Journal of Antimicrobial Chemotherapy* 44 (August 1999): 251–61.

111 One remarkable story came from Orlando, Florida: "County Lines Cancer Patient Refuses to Lose Her Zest for Life," by Jim Toner, *Orlando Sentinel,* July 15, 2001.

111 In one typical trial study of 200 patients: "New Antibiotics to Fight Superbug," *Independent,* September 28, 1998, 10.

111 In an MRSA trial of 460 patients: "FDA Advisers Recommend Approval of Linezolid," *Reuters Health,* March 27, 2000.

111 In a study of forty-four patients: ibid.

111 One of the many doctors: *Independent,* September 28, 1998.

113 Zyvox, by contrast, cost about $140 a day: "FDA Approves New Drug to Attack Resistant Germs," by Sheryl Gay Stolberg, *New York Times,* April 19, 2000, sec. A, p. 19.

114 Pharmacia & Upjohn had dutifully reported: "Linezolid: First of a New Drug Class for Gram-Positive Infections," *Drug & Therapeutic Perspectives* 17, no. 9 (2001): 1–6.

114 Quinn used Zyvox on five patients: "Infections Due to Vancomycin-Resistant *Enterococcus faecium* Resistant to Linezolid," by Ronald D. Gonzales et al., *Lancet,* April 14, 2001.

114 Dr. Cameron Durrant: Adam Marcus, "Resistance Found to Potent New Antibiotic," *HealthScout* (a publication of MDchoice.com at www.healthscout.com).

115 In a follow-up study to Synercid trials, RPR had duly noted: "Characterization of Isolates Associated with Emerging Resistance to Quinupristin/Dalfopristin (Synercid) During a Worldwide Clinical Program," by M. Dowzicky et al., *Diagnostic Microbiology and Infectious Disease* 37 (2000): 57–62.

115 In Michigan, however, a more ominous study: "Antimicrobial Resistance in Enterococci Isolated from Turkey Flocks Fed Virginiamycin," by L. A. Welton et al., *Antimicrobial Agents and Chemotherapy* 42, no. 3 (March 1998): 705–8.

7: A DEADLY THREAT IN LIVESTOCK

This chapter is based chiefly on author interviews with Marcus Zervos, M.D., Fred Angulo, D.V.M., Ph.D., Richard Carnevale, D.V.M., and Stephen Sundlof, D.V.M., Ph.D.

117 he spent a whole year studying turkey feces: "Antimicrobial Resistance in Enterococci Isolated from Turkey Flocks Fed Virginiamycin," by Welton et al., *Antimicrobial Agents and Chemotherapy* 42, no. 3 (March 1998): 705–8.

118 In 1996, he and his researchers had gone to test fecal droppings: "Survey for Multiply Resistant Enterococci from Livestock and Associated Feed," by S. White, S. Qaiyumi, R. J. Johnson, and R. S. Schwalbe, presented at the Ninety-Seventh American Society of Microbiology General Meeting, May 1997.

119 CDC investigators purchased chickens from twenty-six grocery stores: author interview with Angulo.

122 America supported a livestock population of 7.5 billion chickens, 293 million turkeys: Online fact sheet at the Animal Health Institute Web Site, www.ahi.org.

122 A pig, for example, received an average of ten different antibiotics: from a report on the Soil Association on the British Broadcasting Corporation, October 8, 1999.

123 Growth promoters, usually mixed in with food: "Hogging It," a report by the Union of Concerned Scientists, as recounted in the UCS magazine *Nucleus,* Spring 2001.

123 American farmers used seventeen different antibiotics on their livestock: Center for Science in the Public Interest.

124 In September 1999, a three-year-old girl had died: "An Outbreak of Bacteria Kills a Child," *New York Times,* September 6, 1999, sec. B, p. 5.

124 The following summer, another three-year-old girl died: *Reuters Health,* July 31, 2000.

124 North of the border in Walkerton, Ontario: *Reuters,* May 31, 2000.

124 the CDC estimated that 73,000 Americans incurred diarrhea each year: "The Antimicrobial Resistance Patterns of *Escherichia coli* 0157:H7—NARMS, 1996–1999," by K. Johnson et al., a report from the National Antimicrobial Resistance Monitoring System to the International Conference on Emerging Infectious Diseases, 2000.

125 it accounted for some 2.4 million infections each year: author interview with Angulo.

125 About 100 Americans died each year from *Campylobacter:* author interview with Angulo.

125 Rare in the United States a generation ago, it still affected fewer: author interview with Angulo.

126 Together, *Campylobacter* and *Salmonella* accounted for 80 percent: Centers for Disease Control.

129 Ashley Mulroy, of Wheeling, West Virginia, read a shocking story: "Teenager's Science Project Reveals Issues About the Water Some People Drink," *CBS Evening News,* February 23, 2001.

8: REVOLUTION IN EUROPE

This chapter is based chiefly on author interviews with Henrik C. Wegener, C.V.M., Ph.D., Christina Greko, C.V.M., Fred Angulo, C.V.M., Ph.D., and Kirk Smith, D.V.M., Ph.D.

143 retail prices for meat stayed about the same: "The Consequences of Discontinued Use of Antimicrobial Growth Promoters (AGPs) for Food Producing Animals in Denmark," by H. C. Wegener et al., International Conference on Emerging Infectious Diseases 2000, Abstract: Antimicrobial Resistance, 2.

143 "I'm sure VRE can transfer from animals to people": author interview with Richard Carnevale.

144 Sundlof had been advised by CVM's chief counsel: author interview with Stephen Sundlof.

145 "New animal drugs," he would explain: ibid.

147 In all, Smith and his team looked at 6,674 human cases: Quinolone Resistant *Campylobacter jejuni* Infections in Minnesota, 1992–1998," Kirk E. Smith et al., *New England Journal of Medicine* 340 (May 20, 1999): 1525–32.

149 In 1997, just one year after the introduction of fluoroquinolones: author interview with Fred Angulo.

150 On June 18, 1998, Henrik Wegener and his colleagues: "An Outbreak of Multidrug-Resistant, Quinolone-Resistant Salmonella Enterica Scrotype Typhimurium DT104," by Kare Molbak et al., *New England Journal of Medicine* 341, no. 19 (November 4, 1999): 1420–25.

152 a British researcher named John Threlfall: "Increasing Spectrum of Resistance in Multiresistant *Salmonella typhimurium,*" by E. J. Threlfall et al., *Lancet* 347 (1996): 1053–54. Also, author interview with John Threlfall.

152 By 1996, 16 percent of the DT104 in animals was quinolone resistant: "Protecting the Crown Jewels of Medicine," an overview published by the Center for Science in the Public Interest, Washington, D.C., 1998.

152 But the percentage of DT104 in people: author interview with Fred Angulo; also, "Emergence of Multidrug-Resistant *Salmonella enterica* Serotype Typhimurium DT104 Infections in the United States," by M. Kathleen Glynn et al., *New England Journal of Medicine* 338, no. 19 (May 7, 1998): 1333–38.

153 A first documented outbreak in the United States had arisen: Reported by G. Hosek, et al., "Multidrug-Resistant Salmonella Serotype Typhimurium—United States," 1996, *MMWR* 46, no. 14 (April 11, 1997): 308–10.

153 The next U.S. outbreak was both more widespread and more easily traced: "Two Outbreaks of Multidrug-Resistant *Salmonella* Typhimurium DT104 Infections Linked to Raw-Milk Cheese in Northern California," by Sara H. Cody et al., *Journal of the American Medical Association* 281, no. 19 (May 19, 1999): 1805–15.

154 For Cynthia Hawley, forty-five, a brutal acquaintance with DT104 began: "Outbreak," by Amanda Spake, *U.S. News & World Report,* November 24, 1997, 71.

155 By 1998, Fred Angulo and colleagues at the CDC had determined: "Emerging Quinolone-Resistant Salmonella in the United States," Hallgeir Herikstag et al., *Emerging Infectious Diseases* 3, no. 3 (1997): 371–72.

156 In the Philippines: ibid.

156 "We have gotten away from talking about zero risk": author interview with several participants.

157 In Oregon, a patient who appeared to have incurred this infection: "Emergence of Fluoroquinolone-Resistant *Salmonella* Infections in the United States: Nosocomial Outbreaks Suggest a Changing Epidemiology," by S. Olsen et al., National Antimicrobial Resistance Monitoring System, Presentations, 2000, the Centers for Disease Control.

9: BREAKOUT

This chapter is based chiefly on author interviews with Mona Tice, Tim Naimi, M.D., M.P.H., and Patrick Schlievert, Ph.D.

162 Dr. Belani murmured that she would do her best: author interview with Dr. Kiran Belani.

169 a fourth child succumbed: "Four Pediatric Deaths from Community-Acquired Methicillin-Resistant *Staphylococcus aureus*—Minnesota and North Dakota, 1997–1999," by C. Hunt et al., *Morbidity and Mortality Weekly Review* 48, no. 32 (August 20, 1999): 707–10.

169 A short walk from Naimi's office: author interview with Patrick Schlievert. See also "Exotoxins of *Staphylococcus aureus*," by Martin M. Dinges, Paul M. Orwin, and Patrick M. Schlievert, *Clinical Microbiology Reviews* 13, no. 1 (January 2000): 16–34.

170 But Schlievert's bold explanation of toxic shock syndrome: A thorough account of Schlievert's investigation of TSS is found in *The Coming Plague*, by Laurie Garrett (Penguin Books, 1994), 402–5.

10: THE OLD MAN'S FRIEND

This chapter is based chiefly on interviews with Susan Donelan, M.D., Peter Applebaum, M.D., Michael Jacobs, M.D., and Alexander Tomasz, Ph.D.

173 it accounted for the lion's share of 500,000 cases: statement by Anthony S. Fauci, M.D., director of the National Institute of Allergy and Infectious Diseases, National Institutes of Health, before the Senate Committee on Health, Education, Labor, and Pensions Subcommittee on Public Health and Safety, February 25, 1999.

174 Globally, *S. pneumo* was estimated: ibid.

174 Indeed, 70 percent of all respiratory infections *were* viral: author interview with Don Low, M.D.

175 Among hospital patients who had contracted pneumonia: "Penicillin-Resistant Pneumococci: New-Generation Fluoroquinolones and Lower Respiratory Tract Infections," by Julio Ramirez, M.D., *Hospital Medicine* 35, no. 6 (1999): 43–49.

175 Up to 61 percent of the children in some daycare centers: ibid.

177 One day in the spring of 1977, Peter Applebaum: author interview with Applebaum. See also Applebaum's "*Streptococcus pneumoniae* Resistant to Penicillin and Chloramphenicol," *Lancet,* November 12, 1977, 995–97. Michael Jacobs's account of the South Africa outbreak is reported, with colleagues, as "Emergence of Multiply Resistant Pneumococci," *New England Journal of Medicine* 299, no. 14 (October 5, 1978): 735–40.

180 He had come as a young man to America from Hungary in 1956: author interview with Alexander Tomasz.

182 Soon de Lencastre realized there were two closely related clones: author interview with Lencastre. Also, see "Carriage of Internationally Spread Clones of *Streptococcus pneumoniae* with Unusual Drug Resistance Patterns in Children Attending Day Care Centers in Lisbon, Portugal," by Raquel Sa-Leao, Alexander Tomasz, Ilda Santo Sanches, Antonio Brito-Avo, Sigurdur E. Vilhelmsson, Karl G. Kristinsson, and Herminia de Lencastre, *Journal of Infectious Diseases* 182 (October 2000): 1153–60.

182 Iceland, where strict government health protocols: "Evidence for the Introduction of a Multiresistant Clone of Serotype 6B *Streptococcus pneumoniae* from Spain to Iceland in the Late 1980s," by Sonia Soares, Karl G. Kristinsson, James M. Musser, and Alexander Tomasz, *Journal of Infectious Diseases* 168 (July 1993): 158–63.

183 A national survey by the CDC from 1979 to 1986: discussed in "Antimicrobial Resistance in *Streptococcus pneumoniae*: Implications for Treatment in the New Century," by Joseph P. Lynch III and Fernando J. Martinez, M.D., C.M.E., the Medical Education Collaborative, Ortho-McNeil.

183 Then, in 1993, reports began to appear: "Drug-Resistant *Streptococcus pneumoniae*—Kentucky and Tennessee, 1993," *Morbidity and Mortality Report* 43, no. 2 (January 21, 1994): 23–25, 31. See also "Prevalence of Penicillin-Resistant *Streptococcus pneumoniae*—Connecticut, 1992–1993," *MMWR* 43, no. 12 (April 1, 1994): 216–17, 223.

183 The CDC would determine that by the late 1990s: "Increasing Prevalence of Multidrug-Resistant *Streptococcus pneumoniae* in the United States," by Cynthia G. Whitney et al., *New England Journal of Medicine* 343, no. 26 (December 28, 2000): 1917–24.

183 But Don Low, who had fought so hard: "Decreased Susceptibility of *Streptococcus pneumoniae* to Fluoroquinolones in Canada," by Danny K. Chen, Allison McGeer, Joyce C. de Azavedo, and Donald E. Low, *New England Journal of Medicine* 341, no. 4 (July 22, 1999): 233–39.

184 Clinical isolates of *S. pneumo* had been identified: "Mechanisms of Tolerance to Vancomycin in *Streptococcus pneumoniae*," by Robyn M. Atkinson et al., *Infectious Medicine* 17, no. 12 (2000): 793–801.

184 Bill Jarvis of the CDC assumed: author interview with Jarvis.

185 Perhaps, as Tomasz mused: author interview with Tomasz.

11: FLESHEATERS

This chapter was drawn chiefly from author interviews with Don Low, M.D., Dennis Stevens, M.D., and Vince Fischetti, Ph.D. General facts on necrotizing fasciitis can be found through the National Necrotizing Fasciitis Foundation (dbatdorff@aol.com).

187 For Evangeline Ames Murray: author interview with Murray.

190 That, apparently, was the case with George Poste: author interview with Poste.

192 Dr. Dennis Stevens, an Idaho-based infectious diseases specialist: author interview with Stevens.

194 If Stevens and Schlievert's paper: "Severe Group A Streptococcal Infections Associated with a Toxic Shock-like Syndrome and Scarlet Fever Toxin A," by D. L. Stevens et al., *New England Journal of Medicine* 321, no. 1 (July 6, 1989): 1–7.

194 Henson, fifty-three, was at the pinnacle of an extraordinary career: *Entertainment Weekly,* May 16, 1997, 132.

195 On the coffee table of his hospital office: Two of Don Low's seminal papers on Group A strep infections are "Invasive Group A Streptococcal Infections in Ontario, Canada," by H. Dele Davies and others including Low, *New England Journal of Medicine* 335 (August 22, 1996): 547–54; and "Clinical Experience with 20 Cases of Group A Streptococcus Necrotizing Fasciitis and Myonecrosis: 1995 to 1997," by Catherine T. Haywood and others including Low, *Plastic and Reconstructive Surgery* 103, no. 6 (May 1999).

195 One Monday in the fall of 1994, Bouchard was admitted: "The Fight of His Life," by Barry Came, *Maclean's* 107 (December 12, 1994).

197 In Queens, New York, an eight-year-old boy died of it: *New York Times,* April 5, 1995, sec. B, p. 1.

197 In Chicago, a seventy-year-old: *Chicago Sun Times,* December 1, 1999, 72.

197 That same year, a sixty-one-year-old Cuban American man: *Miami Herald,* July 1, 1999, sec. B, p. 1.

197 In San Francisco, also in 1999: *San Francisco Chronicle,* June 18, 1999, sec. A, p. 21.

197 "When a person is admitted": author interview with Dr. Thomas Aragon.

197 Richard Novick, one of the most distinguished researchers: author interview with Novick.

198 Possibly as a harbinger of trouble to come: "Outbreak of Drug-Resistant Strep Bacteria," by Laurie Tarkan, *New York Times,* April 18, 2002, sec. A, p. 23; "Erythromycin-Resistant Group A Streptococci in Schoolchildren in Pittsburgh," by Judith M. Martin, M.D. et al., *New England Journal of Medicine* 346, no. 16 (April 18, 2002): 1200–06.

12: MORE BAD NEWS

This chapter is based chiefly on author interviews with James Rahal, M.D., and his staff at New York Hospital Queens.

204 Unfortunately, he began seeing resistance to the drug: "Nosocomial Outbreak of *Klebsiella* Infection Resistant to Late-Generation Cephalosporins," by Kenneth S. Meyer et al., *Annals of Internal Medicine* 119, no. 5 (September 1, 1993): 353–57.

204 Within two years, Rahal had an epidemic of ceftazidime-resistant *Klebsiella:* author interview with Rahal. Also, see "Identification of TEM-26 B-lactamase Responsible for a Major Outbreak of Ceftazidime-Resistant *Klebsiella pneumoniae,*" by Carl Urban et al., *Antimicrobial Agents and Chemotherapy* 38, no. 2 (February 1994): 392–95.

204 He'd just seen his first *Klebsiella* resistant to imipenem: "Clinical Characteristics and Molecular Epidemiology Associated with Imipenem-Resistant *Klebsiella pneumoniae,*" by Muhammad Ahmad et al., *Clinical Infectious Diseases* 29 (August 1999): 352–55.

205 The answer was to stop using ceftazidime and all other cephalosporins: "Class Restriction of Cephalosporin Use to Control Total Cephalosporin Resistance in Nosocomial *Klebsiella,*" by James J. Rahal et al., *Journal of the American Medical Association* 280, no. 14 (October 14, 1998): 1233–37.

206 a pharmaceutical nightmare called polymyxin: "Clinical and Molecular Epidemiology of Acinetobacter Infections Sensitive Only to Polymyxin B and Sulbactam," by Eddie S. Go et al., *Lancet* 344 (November 12, 1994): 1329–32.

207 They found that 44 percent of the *Klebsiella pneumo* isolates: "Antimicrobial Resistance in Enterobacteriaceae in Brooklyn, NY: Epidemiology and Relation to Antibiotic Usage Patterns," by Guillermo Saurina et al., *Journal of Antimicrobial Chemotherapy* 45 (2000): 895–98.

207 *Acinetobacter's* story was particularly chilling: "Endemic Carbapenem-Resistant *Acinetobacter* Species in Brooklyn, New York: Citywide Prevalence, Interinstitutional Spread, and Relation to Antibiotic Usage," by Vivek M. Manikal et al., *Clinical Infectious Diseases* 31 (July 2000): 101–6.

207 In Zurich, Switzerland: Drs. R. Fleisch and Christian Ruef of University Hospital of Zurich, reporting to the fortieth annual Interscience Conference on Antimicrobial Agents and Chemotherapy, hosted by the American Society of Microbiology, September 17–20, 2000, Metro Toronto Convention Centre, Ontario, Canada.

208 In the fall of 2001: author interview with Louis Rice, M.D. Also see "Ceftazidime-Resistant *Klebsiella pneumoniae* Isolates Recovered at the Cleveland Department of Veterans Affairs Medical Center," by Louise B. Rice, Elizabeth C. Eckstein, Jerome DeVente, and David M. Shlaes, *Clinical Infectious Diseases* 23 (July 23, 1996): 118–24.

208 At Brown University's Miriam Hospital in Providence: author interview with Antone Medeiros, M.D.

208 At the Lahey Clinic in Burlington, Massachusetts: author interview with George Jacoby, M.D.

208 At the Baltimore VA medical center: author interview with Judith Johnson.

209 At nearby Kingsbrook Jewish Medical Center: author interview with Steven E. Brooks, M.D. Also see "Are We Doing Enough to Contain *Acinetobacter* Infections?" by Steven E. Brooks, Ph.D., letter to the editor, *Infection Control and Hospital Epidemiology,* May 2000, 304.

210 In a study published in the fall of 2001: "Widespread Distribution of Urinary Tract Infections Caused by a Multidrug-Resistant *Escherichia coli* Clonal Group," by Amee R. Manges, et al., *New England Journal of Medicine* 345, no. 14 (October 4, 2001): 1007–13.

211 Here, in 1995, Marc Galimand: "Multidrug Resistance in *Yersinia pestis* Mediated by a Transferable Plasmid," by Marc Galimand et al., *New England Journal of Medicine* 337, no. 10 (September 4, 1997): 677–80.

212 In the roughly 2,000 years since it had appeared as a human pathogen: A succinct history of plague appears in Laurie Garrett's *The Coming Plague* (Penguin, 1994), 237–39.

214 But according to Dr. Ken Alibek: See *Biohazard: The Chilling True Story of the Largest Covert Biological Weapons Program in the World—Told from Inside by the Man Who Ran It,* by Ken Alibek with Stephen Handelman (Random House, 1999).

215 In October 2001, for example: Jaime R. Torres, CNN Health, October 11, 2001.

215 Four years after their initial finding: "Transferable Plasmid-Mediated Resistance to Streptomycin in Clinical Isolate of *Yersinia pestis,*" by Annie Guiyoule ct al., *Emerging Infectious Diseases* 7, no. 1 (2001), 43–48.

13: HOPE IN FROGS AND DRAGONS

The story of Magainin is based chiefly on author interviews with Michael Zasloff. Also helpful was Zasloff's unpublished account of the magainin story, "The Commercial Development of the Antimicrobial Peptide Pexigainin," January 2000. One of the best general-audience overviews of peptides is "Ancient System Gets New Respect," by Trisha Gura, Science 291, no. 5511, 2068. The account of searching for peptides in Komodo dragons is based chiefly on author interviews with Terry Fredeking, Jon Arnett, and Don Gillespie.

218 So promising was Zasloff's finding: "Magainins, a Class of Antimicrobial Peptides from *Xenopus* Skin: Isolation, Characterization of Two Active Forms, and Partial cDNA Sequence of a Precursor," by Michael Zasloff, *Proceedings of the National Academy of Science, USA* 84 (August 1987): 5449–53.

218 "If only part of their laboratory promise is fulfilled": Editorial, *New York Times,* July 31, 1987.

219 In 1981, a Swedish researcher named Hans Boman: H. Steiner et al., *Nature* 292, pp. 246–48. See also Boman's review "Antibacterial Peptides: Key Components Needed in Immunity," *Cell* 65 (April 19, 1991): 205–7.

219 In the 1960s, a researcher at New York's Rockefeller University: M. C. Modrzakowski and J. K. Spitznagel, *Infect. Immun.* 25 (1979): 597–602.

220 A decade later, Robert Lehrer: author interview with Robert Lehrer.

223 Robert E. W. Hancock: For starters, see "Peptide Antibiotics," a review by Robert E. W. Hancock, *Lancet* 349, no. 9049 (February 8, 1997).

223 John Forrest, a professor at Yale: "Biotech Discovers the Shark," by Judith Masslo Anderson, *MD,* October 1993, 43–55.

224 Eventually, Zasloff found a way to purify shark squalamine: "Squalamine: An Aminosterol Antibiotic from the Shark," by Karen S. Moore et al., *Proceedings of the National Academy of Sciences USA* 90 (February 1993): 1354–58.

226 The panel, composed of seven experts: Zasloff's unpublished account, "The Commercial Development of the Antimicrobial Peptide Pexigainin," January 2000.

14: BACTERIA BUSTERS

The account of Felix d'Herelle's life is drawn in part from William C. Summers's scholarly biography, Felix d'Herelle and the Origins of Molecular Biology *(Yale University Press, 1999). Many details about George Eliava and the Eliava Institute are drawn from interviews with the institute's Nina Chanishvili.*

234 On a frigid day in January 2001: author interview with Alfred Gertler.

237 Phages are viruses: Good factual descriptions of phages can be found in "Phage Therapy: Past History and Future Prospects," by Richard M. Carlton, *Archivum Immunologiae et Therapiae Experimentalis* 47 (1999): 267–74; "The Return of the Phage," by Julie Wakefield, *Smithsonian,* October 2000, 43–46; "Return of a Killer," by Brendan I. Koerner, *U.S. News & World Report,* November 2, 1998; and "The Good Virus," by Peter Radetsky, *Discover,* November 1996.

238 Born in 1873 in Montreal: This and subsequent details of d'Herelle's early life are in Summers's *Felix d'Herelle.*

241 In a crisp, two-page paper that startled the scientific world: *"Sur un microbe invisible antagoniste des bacilles dysenteriques,"* by Felix d'Herelle. *Comptes rendus Acad. Sciences* 1917: 373–75.

243 Eliava was as brash and dramatic a character: Many details of Eliava's life and career are drawn from author interviews with Nina Chanishvili of the Eliava Institute.

248 D'Herelle and his wife arrived by ship in October 1933: This and other details of d'Herelle's time in Russia are drawn from Summers's *Felix d'Herelle,* as well as "Felix d'Herelle in Russia," by D. P. Shrayer, *Bulletin of the Institut Pasteur* 94 (1996): 91–96.

250 An article in the *Journal of the American Medical Association: Journal of the American Medical Association* 100, no. 3 (1933): 110–13.

250 One product called Enterofagos: Dr. Paul Barrow in "The Virus That Cures," BBC.

251 Early preparations were often impure: "Phage Therapy," by Richard Carlton, *Archivum Immunologiae et Therapiae Experimentalis* 47 (1999): 267–74.

252 Crucial work was also done: "Results of Bacteriophage Treatment of Suppurative Bacterial Infection in the Years 1981–1986," by Stefan Slopek et al., *Archivum Immunologiae et Therapiae Experimentalis* 35 (1987): 569–83. See also an appraisal of Slopek's work in "Bacteriophages Show Promise As Antimicrobial Agents," by J. Alisky et al., *Journal of Infection* 36 (1998): 5–15.

253 Beginning in 1982, British researchers: "Successful Treatment of Experimental *Escherichia coli* Infections in Mice Using Phage: Its General Superiority over

Antibiotics," by H. Williams Smith and M. B. Huggins, *Journal of General Microbiology* 1238 (1982): 307–18. Also, "The Control of Experimental *Escherichia coli* Diarrhoea in Calves by Means of Bacteriophages," by H. Williams Smith, Michael B. Huggins, and Kathleen M. Shaw, *Journal of General Microbiology* 133 (1987): 1111–26.

253 To researchers who had worked with George Eliava: author interview with Nina Chanishvili.

254 Dr. Nina Chanishvili and her colleagues would hurry: Along with author interviews with Dr. Chanishvili, some details about the institute's recent history are drawn from "A Stalinist Antibiotic Alternative," by Lawrence Osborne, *New York Times Magazine,* February 6, 2000, 50–55.

256 Carl Merril had learned of bacteriophages: author interview with Merril.

256 In 1945, Salvador Luria and Max Delbruck had studied: "1969: Max Delbruck, Salvador Luria and Alfred Hershey," a Nobel chronicle by Tonsei N.K. Raju, *Lancet* 354, no. 9180 (August 28, 1999), 784.

259 With that, Carlton staked $50,000: author interview with Carlton.

259 Merril, Adhya, and Carlton published their findings on serial passage: "Long-Circulating Bacteriophage As Antibacterial Agents," by Carl R. Merril et al., *Proceedings of the National Academy of Sciences USA* 93 (April 1996): 3188–92.

260 In a signed commentary: "Smaller Fleas: Ad Infinitum," by Joshua Lederberg in ibid, 3167–68.

260 "The Good Virus," by medical writer Peter Radetsky: *Discover,* November 1996.

260 Two respected scientists: "Phage Therapy Revisited: The Population Biology of a Bacterial Infection and Its Treatment with Bacteriophage and Antibiotics," by Bruce R. Levin and J. J. Bull, *American Naturalist* 147, no. 6 (June 1996).

261 He had made millions by licensing: author interview with Caisey Harlingten.

262 Eventually the Georgians would use an intravenous phage: author interview with Nina Chanishvili.

263 Honour explained that he'd just spent: author interview with Richard Honour.

265 "We gave the Americans access to all this background research": Lawrence Osborne, "A Stalinist Antibiotic Alternative," *New York Times Magazine,* February 6, 2000, 50.

266 Soon, however, this potential alliance fizzled: author interview with Sandro Sulakvelidze.

268 Honour got a desperate call: author interview with Richard Honour. Also, "Defeat of a Superbug?" ABC News.com, September 16, 1999.

15: PEERING INTO THE ABYSS

274 Historically, the Soviet Union had controlled TB with ruthless effectiveness: Abigail Zuger, "Russia Has Few Weapons as Infectious Diseases Surge," *New York Times,* December 5, 2000, F1.

275 Each year, 300,000 new prisoners were incarcerated: Yevgenia Borisova, "Crossing Borders" and "Kemerovo: One Region's TB Profile," *Moscow Times,* January 29, 2000. Also, *New York Times,* December 5, 2000.

275 Paul Farmer, a forty-one-year-old Harvard-trained doctor: author interviews with Farmer. Also, see "The Good Doctor," by Tracy Kidder, *The New Yorker,* July 10, 2000, 40–58.

276 One day in the fall of 1998: John Donnelly and Dave Montgomery, "TB from Ex-Soviet States Resists Most Drugs," *Arizona Republic,* March 21, 1999, sec. A, p. 27.

277 one new study suggested it was actually lowering hospital costs: *Reuters Health,* July 10, 2001.

277 In July 2001, an eighty-five-year-old man in London: "Linezolid Resistance in a Clinical Isolate of *S. aureus,*" by S. Tsiodras et al., *Lancet* 358 (2001): 207–8.

277 In December 2001, researchers at Minnesota's Mayo Clinic reported: A presentation by Immaculada Herrero and colleagues at the Forty-first Interscience Conference on Antimicrobial Agents and Chemotherapy, December 17, 2001.

278 "It's the most cidal drug anyone's every seen": author interview with Frank Talley.

278 Eli Lilly was out. . . . So were Roche and DuPont: author interviews with Chuck Ford and Steve Projan.

278 That fall, a group of researchers from Tufts concluded: *New York Times,* December 1, 2001, sec. C, p. 1.

281 "The world of Synercid": author interview with François Bompart.

281 The other boost to the field: *New York Times,* July 25, 2001, sec. C, p. 17.

282 In February 1997: author interviews with William Jarvis and Steve Quirk. Also see "Control of Vancomycin-Resistant Enterococcus in Health Care Facilities in a Region," by Belinda E. Ostrowsky et al., *New England Journal of Medicine* 344, no. 19 (May 10, 2001): 1427–33.

283 "Hey, what if John Paul Jones": author interview with Barry Farr.

284 The oft-cited case was Communist Hungary: author interview with Marc Lappe.

284 Immunologist Marc Lappe agreed: author interview with Lappe.

284 But Bruce Levin, a prominent population biologist: author interview with Levin.

285 In October 2001, Fred Angulo and Henrik Wegener both contributed: *New England Journal of Medicine* 345, no. 16 (October 18, 2001).

285 Angulo and colleagues tested chicken carcasses: "Quinupristin-Dalfopristin-Resistant *Enterococcus faecium* on Chicken and in Human Stool Specimens," by L. C. McDonald, et al., *New England Journal of Medicine* 345, no. 16 (October 18, 2001): 1155–60.

285 Wegener and other researchers: "Transient Intestinal Carriage After Ingestion of Antibiotic-Resistant *Enterococcus faecium* from Chicken and Pork," by T. L. Sorenson et al., *New England Journal of Medicine* 345, no. 16 (October 18, 2001): 1161–66.

285 Very possibly, the growing public clamor: *New York Times,* February 10, 2002, Sec. A, p. 1.

286 with profits of about $1.6 billion: *New York Times,* November 13, 2001, sec. B, p. 6.

286 Richard Levins, a population biologist at Harvard: author interview with Levins.

A SELECTED BIBLIOGRAPHY

Alcamo, Edward. *Fundamentals of Microbiology*. 4th Ed. New York: Benjamin/ Cummings Publishing, 1994. A marvelously readable college textbook.

Cannon, Geoffrey. *Superbug*. London: Virgin Publishing, 1995. A ruminative and compelling overview of antibiotic resistance.

de Kruif, Paul. *Microbe Hunters*. New York: Harcourt Brace/Harvest, 1954. Originally published in 1926, this classic compendium of stories of microbe hunters before the antibiotic era inspired countless teenagers to become scientists. Still great fun.

Fisher, Jeffrey A. *The Plague Makers*. New York: Simon & Schuster, 1994. A handy overview of the history of antibiotics—and of antibiotic resistance.

Garrett, Laurie. *Betrayal of Trust*. New York: Hyperion, 2000. Fascinating investigations of recent outbreaks of plague, Ebola, multidrug-resistant tuberculosis, and more.

———. *The Coming Plague*. New York: Penguin, 1994. A definitive, Pulitzer Prize–winning overview of global infectious disease threats, including drug-resistant bacteria.

Lappe, Marc. *When Antibiotics Fail*. Berkeley, Calif.: North Atlantic Books, 1995. A prescient and passionate work first published in 1986.

Levy, Stuart. *The Antibiotic Paradox*. New York: Plenum Press, 1992. Though now a decade old, Levy's elegantly succinct work remains the seminal study of antibiotic resistance. It explains the science of the subject in considerable detail, yet remains a book for the layperson, readable and compelling. Available through Levy's Alliance for the Prudent Use of Antibiotics (see Web Sites on page 317).

Miller, Judith, Stephen Engelberg, and William Broad. *Germs*. New York: Simon & Schuster, 2001. A fascinating history of germ warfare in the United States and the U.S.S.R. that discusses the engineering of drug-resistant plague.

Osterholm, Michael. *Living Terrors.* New York: Delacorte Press, 2001. Another fascinating look at germ warfare from the well-known former Minnesota state epidemiologist.

Schell, Orville. *Modern Meat.* New York: Random House, 1984. This ground-breaking work was the first to report on the history and consequences of growth promoters in agriculture.

Summers, William C. *Felix d'Herelle and the Origins of Molecular Biology.* New Haven, Conn.: Yale University Press, 1999. A painstaking history of the great phage pioneer and his work.

Wainwright, Milton. *Miracle Cure.* Oxford: Basil Blackwell, 1990. An entertaining and cogent history of the development of penicillin and other essential antibiotics.

WEB SITES

Several Web sites now offer a wealth of information on antibiotic resistance, including recommendations for doctors and patients. The first two listings are perhaps the most comprehensive; the others follow in no particular order.

www.keepantibioticsworking.com—Several consumer groups, among them Environmental Defense, the Union of Concerned Scientists, and the Sierra Club, have banded together to make this a one-stop-shopping site for all the latest news and links on antibiotic resistance. An amazing resource that covers all aspects of the subject but focuses on agricultural antibiotic use.

www.apua.com—This is the site for Stuart Levy's Alliance for the Prudent Use of Antibiotics, also a cornucopia of news, recommendations, and links.

in.fullcoverage.yahoo.com/fc/India/Antibiotics_and_Microbiology/—the Yahoo site for antibiotics and microbiology. Loads of good stuff.

www.evergreen.edu/phage—the best Web site devoted to bacteriophages, established and maintained by Betty Kutter of Evergreen State College.

www.intralytix.com—the Web site for Glenn Morris's phage start-up, with good general information about phages.

www.expobio.com—the site for Exponential Biotherapies, another phage start-up.

www.phagetx.com—the site for Phage Therapeutics, the third U.S. phage start-up.

www.cdc.gov—This is the opening page of the Centers for Disease Control's Web site, which affords access to a number of other important sites, including the CDC's regular publications, like the *Morbidity and Mortality Weekly Report,* as well as *Emerging Infectious Diseases.* The site has a section on drug-resistant bacteria.

www.public.iastate.edu/~fuchs/abr/antib.html—a fascinating and quirky compendium of articles on antibiotic resistance.

www.virology.net/garryfavwebjournals.html—an amazing (and exhaustive) list of scores of microbiology journals, with links to all.

www.scirus.com/?h—an overall medical information Web site, with lots of links for antibiotic resistance.

INDEX

Lederberg, Joshua, 14, 181, 260, 267
Lederle pharmaceutical company, 52
Leeuwenhoek, Anton van, 31
Lehrer, Robert, 220, 223
Leuconostoc bacteria, 59
Levin, Bruce, 260, 284–285
Levins, Richard, 286–288
Levy, Stuart, 14, 18, 26, 54, 55, 133, 282
Lewis, Sinclair: *Arrowsmith,* 247n2
lincomycin, 123, 129
linezolid (Zyvox), 103, 107–115, 144, 184, 200, 236, 271, 277, 278
Locilex, 226, 227
London Public Health Laboratory Service, 152
Low, Don, 48–49, 184, 185, 186, 194–196
Luria, Salvador, 256
Lyme disease spirochete, 26n1
lysozyme, 33, 34

McCarty, Maclyn N., 180
McGeer, Allison, 48–49, 195
MacLeod, Colin M., 180
McLeod, Gavin, 188–189
macrolides, 179, 198–199
Mad Cow disease, 140
magainins (peptides), 218, 220
Magainin (firm), 221–228
Maggelburg University (Germany), 57
Manges, Amee, 210–211
Marfan syndrome, 268
Marine Biological Laboratory (Maine), 223
Martin, Judith M., 198
Massachusetts Institute of Technology, 64
Mayo Clinic (Minnesota), 167, 277
measles, 179
mecA (resistance gene), 39, 66
Medeiros, Antone, 208
Meipariani, Amiran, 255, 265
meningitis, 29, 35, 41; *S. pneumo* and, 11, 174, 177–179; vaccine against, 185
Merck pharmaceuticals, 221, 281
Merril, Carl, 256–260, 266
Metchnikoff, Elie, 242
methicillin, 13, 38, 42, 69, 113; resistance to, 165 (*see also* MRSA)
MIC (minimum inhibitory concentration) of drug, 72–78 *passim,* 84, 86, 92, 99, 151
milk, 154. *See also* food-borne diseases
Miller, George, 272
Miltonberger, Teresa, 92–93

Minnesota: Department of Agriculture, 148–149; Public Health Department, 134, 145, 146, 147, 165
minocycline, 213
Modern Meat (Schell), 52
Moellering, Robert, 99–100, 101, 102, 110, 111–112
Morris, J. Glenn Jr., 3–14 *passim,* 19, 21–22, 118, 126, 130, 184; opposes quinolone use, 132, 156–157, 158; and phages, 237, 253, 263, 266–267, 270–271, 288–289; and VRE, 44–48, 50–51, 55, 100, 262–263, 266, 283
Mount Sinai Hospital (Toronto), 48, 184
MRSA (methicillin-resistant *S. aureus*), 13–14, 38–39, 46, 91, 93, 100; community, 163, 165–172; drug approval lacked, 112–113, 115; effective treatment of, 42–43, 49–50, 80–82, 85, 131, 233, 262–263, 270, 277, (isolation) 87–88, 283–284; first reported, 38–39; MIC of isolates, 75, 76; studies of, 66, 67, 265; Synercid and, 99, 101, 102; as threat, 20, 42, 105; vancomycin resistance added, 70, 71–72, 74, 75–76, (VISA) 76, 77, 86, 92; Zyvox-resistant, 277
Mulroy, Ashley, 129
Murray, Barbara, 139, 226
Mycobacterium, 274
myositis, 190

Nadler, Harriette, 103
nafcillin, 38
Naidoo, J., 62n1
Naimi, Tim, 165–169, 171, 172
nalidixic acid, 146–147, 151, 157
National Institute of Allergy and Infectious Diseases, 174
National Institutes of Health (NIH), 217, 219, 221, 224, 256, 257, 258, 270
National University Hospital (Reykjavik), 183
National Veterinary Institute (Sweden), 137, 138
Native Americans, 163, 165, 167
neomycin, 38
Netherlands, the, 132, 137, 139, 140, 228
New England Journal of Medicine, 62, 194, 199, 285
New Guinea, 178, 180
New Jersey state health department, 84
New York City Public Health Research Institute, 18, 48, 282